DOXXED

How Privacy Abuse Harms

Briony Anderson

With a Foreword by
Mark Wood

BRISTOL
UNIVERSITY
PRESS

First published in Great Britain in 2026 by

Bristol University Press
University of Bristol
1–9 Old Park Hill
Bristol
BS2 8BB
UK
t: +44 (0)117 374 6645
e: bup-info@bristol.ac.uk

Details of international sales and distribution partners are available at bristoluniversitypress.co.uk

DOI: 10.51952/9781529253986

British Library Cataloguing in Publication Data
A catalogue record for this book is available from the British Library

ISBN 978-1-5292-5395-5 hardcover
ISBN 978-1-5292-5397-9 ePub
ISBN 978-1-5292-5398-6 ePdf

Cover design: Lyn Davies Design
Front cover image: Stocksy/kkgas
Bristol University Press uses environmentally responsible print partners.
Printed and bound by CPI Group (UK) Ltd, Croydon, CR0 4YY

Bristol University Press' authorized representative in the European Union is:
Easy Access System Europe, Mustamäe tee 50, 10621 Tallinn, Estonia,
Email: gpsr.requests@easproject.com

FSC
www.fsc.org
MIX
Paper | Supporting responsible forestry
FSC® C013604

Content Warning

This book explores research themes and lived experiences of gender-based violence, family and domestic violence, and digital sexual abuse. Accounts of sexual assault are presented in the words of victim–survivors, which may be confronting and distressing for some readers.

If you or someone you know is experiencing online or offline abuse, the following services provide confidential and free support:

United Kingdom

Victim Support: https://www.victimsupport.org.uk/
Refuge: https://refuge.org.uk/

Australia

E-Safety Commissioner: https://www.esafety.gov.au/
1800RESPECT: 1800 737 732

This book is dedicated to all
victim–survivors of privacy abuse, but
particularly the 18 women and men who
generously shared their experiences with me.

Contents

List of Abbreviations		viii
List of Key Terms		ix
Glossary of Victim–Survivor Experiences		x
About the Author		xiii
Acknowledgements		xiv
Foreword		xv
1	Introduction: How Do Non-Consensual Disclosures Harm?	1
2	Personhood and Place in Online Domains	17
3	The Violence of Non-Consensual Disclosure	31
4	Privacy Harms in the Present and Future	46
5	Overcoming Doxxing: Masculinity and Ruptured Honour	62
6	Resistance in the Aftermath of Doxxing	80
7	Conclusion: Affirming the Right to Informational Autonomy	100
Notes		114
References		115
Index		137

List of Abbreviations

AI	artificial intelligence
DARVO	deny, attack, and reverse victim and offender roles
DFV	domestic and family violence
GBV	gender-based violence
GDPR	the *General Data Protection Regulation* (2016)
HHCRO	the *Harassment, Harmful Communications and Related Offences Act* (2020)
IBSA	image-based sexual abuse
ICT	information and communication technology
LGBTQ+	lesbian, gay, bisexual, trans, and queer, plus other gender and sexuality orientations
NFT	non-fungible token
NPCC	National Police Chiefs' Council
PII	personal, identifying information
TFVA	technology-facilitated violence and abuse

List of Key Terms

coercive control A form of abuse which seeks to control the target through prolonged intimidation, gaslighting, and coercion (see Harris and Woodlock, 2019).

doxxing The non-consensual disclosure of personal, identifying, or sensitive information online, usually with the intention to harass, intimidate, or manipulate the target.

doxtortion The threat or actual release of doxxed information to blackmail or procure financial gain from the target.

hacking The act of illegally accessing a computer, personal device, or other information system (such as a database) with the intention of accessing personal data or other confidential information.

inforgs Any actor, human, or otherwise, that can contain, transmit, and receive information in dialogue with another actor (see Floridi, 2013).

infra-ethics Ethical infrastructure of technologies, where the design of informational computer technologies is not neutral but imbued with the values of their designers.

the onlife Where offline and online contexts become porous and blur into a hybridized experience of terrestrial and digital reality.

phishing A technique of cyberfraud where the target is approached by a seemingly legitimate 'authority' – for example, a government agency, law enforcement, or a financial institution – seeking payment.

posthumanism A theoretical framework which rejects the primacy of humans, and draws attention to the relational assemblages of new climate, technological, and political forces (see Braidotti, 2019a).

upskirting A technique of producing and sharing image-based sexual abuse where the perpetrator takes photographs or video of a person's genital or anal region, without their knowledge or consent (see Thompson, 2020).

Glossary of Victim–Survivor Experiences

The names of the 18 women and men who shared their experiences of doxxing have been pseudonymized throughout the book, to protect their anonymity and privacy. Any identifying information in their accounts has also been removed. Moreover, the women and men who participated in this study were not asked to report demographic data (for example, age, location, gender identity) as a deliberate measure to affirm and respect their autonomy to decide how much they wished to share of their stories and themselves. As such, I adopted the language they used to refer to themselves – for example, 'a woman', 'young man in a heterosexual relationship', but did not query whether they were cisgendered or gender-diverse. The following table outlines the pseudonyms used throughout the book to identify their accounts of privacy abuse, and it provides a brief overview of the type of doxxing they experienced.

Pseudonym	Experience of doxxing	Demographic information (if disclosed)
Odette	Filmed without consent in a nightclub – upskirted* – and doxxed via Facebook and Instagram. Odette's name, home address, and workplace information was also doxxed. Odette experienced reputational and professional harms, resulting in the loss of her job.	Woman in her mid-20s.
Aisha	Experienced retributive image-based doxxing following a breakup with her ex-partner. Sexually sensitive images and videos were doxxed by her ex-partner on Facebook and WhatsApp to 'pay her back' for ending the relationship.	Woman in her late 20s, living in Australia.
Tegan	Experienced targeting doxxing from a former friend, who doxxed her private reproductive history, including details about her abortion, on Facebook. This resulted in significant relational harms with Tegan's religious family and community.	Woman in her early 30s, living in Australia.

Pseudonym	Experience of doxxing	Demographic information (if disclosed)
Tamara	Victim of doxtortion*, where the sale of her sensitive sexual image and personal contact information was threatened. After being harassed by men who had 'purchased' her abuse material, Tamara found out that the doxxer was a close personal friend.	Woman in her early 30s.
Chelsea	Experienced a breach of her identifying documents and personal information in a large-scale telecommunications data hack, resulting in targeted phishing* scams and other efforts to financially defraud her.	Woman in her mid-20s, living in Australia.
Molly	Survived a prolonged, targeted doxxing campaign in which her confidential work materials were hacked, and her personal information and home address were doxxed, in addition to the school location of her young child.	Woman in her late 30s.
Hara	Sensitive sexual video was doxxed after her device was hacked anonymously. Found out that she had been doxxed after a friend asked her about the video, having seen it on social media.	Woman in her early 20s.
Carol	Identifying workplace documents, including confidential materials, were doxxed in a targeted, anonymous doxxing campaign to discredit her professionalism. This resulted in Carol experiencing significant jeopardization of her position at work.	Woman in her 30s.
Jana	Experienced targeted doxxing of her home address and other identifying information from a male colleague, following misogynistic hate speech and harassment online. Jana was successful in bringing a complaint about his behaviour to her workplace, resulting in him being fired.	Woman in her 30s.
Jessie	Personal social media account on Facebook was hacked, and used to disseminate sensitive sexual photos. Jessie was locked out of her account and experienced significant challenges in regaining access, limiting her ability to report her image-based doxxing.	Woman in her 20s, living in Australia.
Sabiyah	Experienced retributive image-based doxxing in the aftermath of breaking up with her ex-partner, who posted her intimate photos on social media. Sabiyah confronted her perpetrator and was successful in having the doxxed photos deleted online.	Woman in her 20s.

Pseudonym	Experience of doxxing	Demographic information (if disclosed)
Eden	Defined doxxing as a process which can be affirming when it allows you to connect with others in the community to share insights and advice, as opposed to harmful when information is shared without consent.	Woman in her early 30s, living in a non-western context (although not specified).
Eila	Experienced image-based abuse when her partner doxxed an intimate video of her after a disagreement. Eila's partner denied sharing it, blaming a friend for the doxxing attack.	Woman in her mid-20s.
Joe	Filmed without knowledge by a family member when caring for a sick animal. The video was shared without his consent by a family friend who shamed and humiliated him within his school peer network.	Man in his early 20s, reflecting on doxxing experienced as a teenager.
David	Experienced self-defined doxxing when allegations of his mistreatment of his partner were made public on social media. Evidence of coercive control*, as he pressured his partner to dispute the allegations of abuse publicly.	Man in his late 20s.
Walter	Experienced self-defined doxxing when a close friend came forward to accuse him of perpetrating rape. Evidence of deny, attack, reverse victim and offender (DARVO*) techniques as he suggested that his friend was motivated to 'lie' about the allegations due to unrequited romantic feelings.	Man in his late 20s.
Ethan	Defined doxxing as a community practice of building trust and credibility in cryptocurrency and non-fungible token (NFT*) forums. While this interview is not drawn on extensively throughout the book, Ethan's experience is briefly discussed in relation to the framework of *informational autonomy* developed in the conclusion.	Man in his late 20s, living in Australia.
Andrew	Filmed without knowledge at a university party while under the influence of alcohol. Andrew was acting violently towards another attendee. The video was doxxed on Facebook and WhatsApp. Andrew believes the video was doxxed by a university peer who wanted to pursue his girlfriend at the time.	Man in his late 20s, currently living in the United States.

* See List of Abbreviations and List of Key Terms for definitions.

About the Author

Briony Anderson is a Career Development Fellow in the Department of Sociology at Durham University, United Kingdom. She is a member of the Centre for Research into Violence and Abuse (CRiVA), where she undertakes research on privacy abuse, gendered-harms, and LGBTQ+ victim–survivor experiences of domestic and family violence. Briony's research has been published in the *British Journal of Criminology, Crime, Media, Culture, Qualitative Inquiry, Criminology & Criminal Justice* and the *International Journal for Crime, Justice, and Social Democracy*. Her research has been recognized at the *International Technology Knowledge and Society* conference, where she received the *Emerging Scholar Award* in València, Spain in 2024.

Acknowledgements

This book is drawn from my doctoral thesis, researched and written on Wurundjeri land in Melbourne, Australia, from 2020 to early 2024. This time period was one of significant social and technological upheaval. From global pandemics, to Trump (back) in the White House, to the explosion of generative artificial intelligence (AI) – our digital and social landscapes continue to be in flux. I am enormously grateful to the 18 women and men I interviewed for providing complex, honest, and vulnerable insights into the harms of doxxing and privacy abuse. In such challenging and changing digital times, their contributions to our understanding of privacy and data rights are vital.

To the wonderful editorial team at Bristol University Press – Rebecca Tomlinson, Zoë Forbes, and Grace Carroll – thank you for your support in developing this book. I am deeply grateful for the mentorship and support of Associate Professor Bianca Fileborn and Dr Dave McDonald for their supervision of my doctoral research which has inspired this book. Thanks also to Professor Alison Young for her mentorship and encouragement. I was privileged to receive funding and professional development from the Department of Criminology at the University of Melbourne, funded by the Australian Commonwealth Research Training Program. Special thanks to Assistant Professor Nikki Rutter for her excellent mentoring, advice, and encouragement.

Finally, I am so grateful to my wonderful community of friends, peers, and colleagues who have enriched my research and learning journey so far – and who continue to inspire me toward work that is activist, feminist, and inclusive. Special thanks to Dr Mark Wood, my long-term collaborator and friend, for providing the foreword to this book. Most importantly, thank you to my family and my partner, Lochlan, for your unwavering support in all things – whether they be international moves, writing books, or some combination of the above.

Foreword

Mark Wood
Deakin University

In our hyperconnected age of Instagram, TikTok, LinkedIn, and the microblogging platform formerly known as Twitter, our lives are increasingly onlife. As the Internet of Things and other 'code spaces' increasingly pervade our lifeworld, we cannot meaningfully speak of time spent 'online' and 'offline' (Floridi et al, 2015; Kitchin and Dodge, 2014; Szakolczai, 2025). We are rarely out of digital earshot from our devices, which continue to listen in and fabricate profiles for platform capitalists. Our digital information is far from a mere shadow of ourselves that, like the protagonist of a Murukami novel, we can detach from by choice. We are our information, and when this part of us is weaponized it can cast a shadow over our security, sense of self, and autonomy. Doxxing, the practice of publishing personal identifying information about someone on the internet, usually with malicious intent, is a quintessentially onlife harm, borne of the internet's information transmission and creation affordances. Yet, despite the growing recognition of doxxing's harms, they remain understudied.

This is just one reason why Briony Anderson's *Doxxed: How Privacy Abuse Harms* is such a necessary and timely book. Through centring the experiences of eighteen victim-survivors of doxxing, *Doxxed* demonstrates why publishing personal identifying information about someone on the internet can represent a harm in its own right. Doxxing is not simply a means to inflict future violence. It is violence: a source of relational rupture, a violation of our information autonomy, and often a deeply gendered form of harassment. It generates not only first-order harms to people's interests and bodily integrity, but second-order harms to their security interests, now and often well into the future (Gross, 1979; Ramsay, 2011).

At the heart of her sophisticated account of doxxing is the concept of informational autonomy. Drawing on Floridi's (2013) information ethics, Briony reframes personal data as not just about a person but of them – integral to identity and deserving of care. When that data is exposed without consent, something deeper than privacy is violated. We are, she argues, informational beings – or inforgs, in Floridi (2013) parlance – composed through and constituted by the data that flows between us and others. Our digital traces are not detachable artefacts; they are entangled with who we are. Doxxing erodes our autonomy over the information that constitutes us; it undermines our agency to control when, and how we share our information, and who we share it with.

The stories in this book make plain what can happen when our informational autonomy is compromised. Doxxing can disrupt families, friendships, and intimate ties; it can reshape how people are seen not only by strangers but by those closest to them. It can expose hidden hierarchies of worth, altering the terrain of trust and safety in both the present and the future. It can leave people feeling shamed, blamed, and unsafe in their own homes. At its most damaging, it severs the thread between who someone is and how they're able to present themselves to the world, producing a kind of erasure that goes beyond reputation.

That's why this book makes a compelling case for a shift away from a deficit model of privacy harms towards an affirmative ethics of informational autonomy. Drawing from Braidotti (2019b), Briony asks us to imagine what freedom from harm could look like – not just the absence of violence, but the presence of care, dignity, and the capacity to choose how and where we participate in social life. Anonymity and privacy must be rethought as active, situated forms of autonomy. They are not about withdrawal but about self-definition across different contexts, with different audiences, on our own terms. They require us to uphold Nissenbaum's (2004) norms of contextual integrity and pay heed to the contexts of information appropriateness and distribution. We must consider 'who is gathering the information, who is analyzing it, who is disseminating it and to whom, the nature of the information, the relationships among the various parties, and even larger institutional and social circumstances' (Nissenbaum, 2004: 154–155). As Briony emphasizes in her manifesto of affirmative information autonomy that closes the book, we must uphold people's right to contextual autonomy – to choose how, when, why, and to whom their information is disclosed – but so too their right to opt in and out of public domains, and their right to respect and care for all information.

Briony approaches all of this with precision and care for the people at the centre of her research. Her book is a major leap forward in our understanding of how doxxing harms, and how people resist these ruptures to their informational autonomy. It is exactly the kind of theoretically sophisticated, multidisciplinary, empirically grounded and innovative research we need to understand the realities of informational harm and the affirmative, caring, and context-sensitive responses they call for.

July 2025

1

Introduction: How Do Non-Consensual Disclosures Harm?

'I wish I was just the only one doxxed. My child was also doxxed.' We were a few minutes into the interview when Molly[1] shared this with me. A single parent in her 30s, Molly is the victim–survivor of a targeted doxxing attack which revealed her personal, identifying, and sensitive information (PII). She endured the non-consensual disclosure of her home address, her personal phone and email contact information, work ID, and other confidential materials on a variety of social media platforms. Molly does not know who perpetrated her doxxing, but believes that they must have been someone close to her at work who had access to her confidential login credentials. The impact of the disclosure of her personal information not only shattered her professional reputation but also threatened her security and safety in her home. Most importantly, the disclosure of her child's identity exacerbated Molly's fear beyond her own personal safety to her child as well.

The consequences of Molly's doxxing included fallout at work, as the disclosure of confidential workplace materials in her care made her look irresponsible and untrustworthy. She nearly lost her job as a result, and endured sanctions and probations in her workplace. However, these consequences of her doxxing were secondary to the ongoing fear that someone would use the doxxed information to approach her, or her child, at their home, school, or workplace. Molly chose to keep her child home from school for many months, fearing the risk of physical violence to her child if she were to leave their home. Reliving her doxxing was not a distant memory or a discrete, linear experience with a defined point of closure. Throughout our interview, the risk of future disclosure of her personal, identifying, and sensitive information lingered heavy and close.

When I asked her about the impact of her child being doxxed, she shared the following:

> One thing I know for sure is that there is a different feeling when you are being doxxed. Being doxxed is like being raped. Most times I feel like for a man to be doxxed, it's better than a woman being doxxed. So I don't know. If my child wasn't involved? I don't know how I would've operated with the [police], if I would've gone ahead with [reporting] it or not.

Molly's experience of doxxing is one of many experiences analysed in this book. Unfortunately, the impact of her doxxing as something that intersects with her identity as a mother, a woman, and a worker is not uncommon. Doxxing represents a violation of information that is indivisible from ourselves, where personal, identifying, and sensitive information is shared in contexts which invite shaming, harassing, and vilifying public responses against the victim–survivor. The information which is leveraged in moments of doxxing is often ordinary and commonplace, such as a full name, a phone number, or a workplace address. However, doxxing exploits ordinary information into enduring, complex, and layered harm experiences which shatters the ability of victim–survivors to live their lives free from harassment, anxiety, and the threat of future disclosure.

This book explores how doxxing produces harms which are experienced in gendered, sensorial, and relational ways. As the accounts of Molly and many others demonstrate, doxxing is a gendered harm which ruptures the free and unimpeded enjoyment of both offline and digital spaces. Doxxing mobilizes forms of harassment, abuse, and violence which sustain and maintain patriarchal systems of domination. These systems of domination undermine the freedom, autonomy, and safety of all victim–survivors, but play out in specific gendered contexts and impacts. This positions women, children, queer, and trans people to be vulnerable to the 'erasure' of their autonomy over personal and sensitive information (Anderson, 2025).

The disclosure of personal, identifying, and sensitive information acts as a vector to shame and condemn victim–survivors. I argue that the violence of doxxing ruptures the autonomy and freedom of victim–survivors to share their personal information in the contexts and to the people of their choice. These ruptures also position victim–survivors along hierarchies of worth maintained and defined by patriarchal value judgements of honour, power, and domination (see Vera-Gray, 2017). Understanding how doxxing harms its victims, and how these harms are enmeshed within the maintenance of patriarchal systems of control and domination, is a key contribution of this book. By contextualizing doxxing as a form of gendered privacy abuse, I grapple with the risks and potentialities of personal information to harm not only in the digital social present but also in the futures to come.

Doxxing in digital social life

Doxxing[2] is a neologism which originally described the practice of 'dropping documents' in hacker subcultures to prove skills and gain acceptance through the disclosure of a target's identity (Douglas, 2016: 200; Anderson and Wood, 2021: 205). Colloquially, doxxing now describes the non-consensual disclosure of personal, identifying, and sensitive information, usually in online contexts such as social media platforms and forum sites. Doxxed information commonly includes identifying information such as home addresses, phone numbers, and places of work, but can also include sensitive data such as intimate photographs, videos, and sound recordings (Huber, 2023). Doxxing shares similarities with other forms of disclosure including outing (Chekola, 1994) and self-identification practices in cryptocurrency and non-fungible token (NFT) forums (Meyns and Dalipi, 2022; Scharfman, 2023) but is distinguished as a practice which leverages the information of another subject without their knowledge or consent.

Doxxing has become synonymous with confrontational and retaliatory uses of social media platforms and digital forums, including 'cancelling' and other mob-style online behaviours where doxxed information is used to harass and intimidate the targeted individual (Wong, 2021; Tandoc et al, 2022). As such, much of the research on doxxing couches it as a form of harassment, a technique of cyberbullying, and a means of extorting and defrauding vulnerable users in digital spaces (McIntyre, 2016; Nadir and Bakhshi, 2018; Dunn, 2021; Foster and Cross, 2024). Technology-facilitated violence literatures have also framed doxxing as an ancillary harm which is used to perpetrate other forms of cybercrime, such as cyberbullying, harassment, and image-based sexual abuse (Chen et al, 2019; Jørgensen and Demant, 2021).

Harassment and abuse stemming from the non-consensual disclosure of personal information is an important starting place in accounting for the harm of doxxing, and is represented in criminological framings of technology-facilitated violence, cyber security, and information and technology studies. However, in this book I depart from conceptualizing doxxing as a means of committing *other* types of harm. Doxxing is itself a complex and enduring harm experience that not only makes victim–survivors vulnerable to physical threats to their security and well-being in the present but places them at risk of ongoing, cyclical abuse of their information in the future. I have elsewhere categorized these harms as 'first-order harms' which can 'interfere with an individual's livelihood and personal life', in addition to setting back their expectations to 'anonymity, the right to a private life and the right to be forgotten' in digital spaces (Anderson and Wood, 2022: 201). 'Second-order harms' are not separate to the first-order harms which arise from doxxing, but 'extend to encompass the prospect of further first-order harms … as well as the emotional and psychological harms resulting from this setback to a targeted individual's security interests' (Anderson and Wood, 2022: 203). Understanding first-order and second-order doxxing harms as complex, interrelated, and ongoing is key to understanding how doxxing represents a unique and complex harm experience independent from other understandings of digital abuse.

Stine Eckert and Jade Metzger-Riftkin (2020: 273) have highlighted that doxxing is experienced as 'a complex, gendered communicative process of harassment'. As I explore in Chapters 3, 4, and 5, doxxing invokes gendered harassment as a means of maintaining patriarchal norms, heteronormativity, and value judgements about what men and women 'should' be. I critique these framings of patriarchal, binary, and heteronormative understandings of gender throughout this book and point out the complex ways that value judgements sustain shame, condemnation, and harassment to victim–survivors of doxxing. With inspiration from Eckert and Metzger-Riftkin's (2020) conceptualization of doxxing as a gendered form of harassment, I add to this framework and argue that the gendered harms that arise from non-consensual disclosure must be identified as a complex form of interpersonal abuse and violence. I consider how non-consensual disclosure invokes gendered dimensions of identity and reputation to sustain harm in ways that reduce the subject's ability to exercise autonomy over their information in both the immediate and enduring aftermath of doxxing. By accounting for the gendered and enduring impacts of doxxing on victim–survivors, I develop a new theoretical and conceptual framing: the need for *informational autonomy*. Informational autonomy illustrates the ways that we are inseparable from our information, where information is not 'about' us but is 'of' us (Floridi, 2011: 556). This framing accounts for the myriad ways that power is invoked through non-consensual disclosure, and is a key contribution arising from this book (Chapters 4 and 5).

Legal remedies for doxxing

To date, there are very few case studies where doxxing has been litigated as a criminal offence. While some jurisdictions – such as the Netherlands,[3] Australia,[4] Indonesia,[5] Hong Kong,[6] New Zealand[7] and South Africa[8] – have explicit laws criminalizing doxxing, many other countries indirectly respond to data and privacy abuse through their legal systems. In the United Kingdom, doxxing is not a criminal offence, but instead may be prosecuted if doxxing leads to harassment,[9] or threatening and abusive messages.[10] Opportunities to report doxxing are also limited in the United Kingdom, with some possible reporting outcomes protected under the *Online Safety Act 2023*, which mandates that platforms remove illegal and harmful content. While global shifts towards legislation which criminalizes doxxing continue to grow, the absence of clear, standardized legal definitions of doxxing makes it difficult for victim–survivors to recognize and advocate for themselves.

Åste Corbridge (2018), an Australian legal scholar, highlights that while existing laws such as the South Australian *Criminal Law Consolidation Act (1935)* can be used to litigate doxxing, they fail to address the complexity and technology-facilitated nature of its perpetration. While doxxing under the *Criminal Law Consolidation Act (1935)* is punishable as a stalking offence in South Australia, Corbridge (2018: 8) identifies that the measure of offence – where a target has had their information disclosed on more than two occasions by the same perpetrator – fails to consider the viral nature of doxxing. As I critiqued with Wood in an earlier publication, 'a victim might receive hundreds of breaches of personally identifiable information, all from different users/perpetrators' (Anderson and Wood, 2021: 220). Presently, there is no mechanism in law to recognize the re-sharing or engagement (such as 'liking' or 'reacting') with doxxed information on social media platforms as a separate offence, even if it further disseminates the doxxed information. Stäcker, Franz and Hett (2025: 12) have recently identified that doxxing events are spurred by viral sharing and harassment campaigns which align individual users to 'act as a cohesive group using internal coordination to guide their collective actions', often using functionalities of the platform – for example, tagging other users, resharing abuse material to new followings – to facilitate doxxing abuse. While legislators are pivoting toward laws which protect PII, the role of technologic affordances in platforming abuse – and the scale of its mediation – remains under-represented in existing laws.

The limitations of legal recognitions of doxxing are not specific to the Australian jurisdictional context. In the United States, the recognition of doxxing as a form of harassment is obscured by competing legal rights to the freedom of speech and expression. Julia MacAllister (2017: 2565) highlights that the right to freedom of speech in the United States should not interfere with the ability to pursue criminal justice remedies for doxxing, as free speech is restricted by the 'true threat' exception where speech creates harm or risk to the target. However, this does not seem to translate into distinct or clear pathways for seeking legal remedies for doxxing in the United States. Like Corbridge, Victoria McIntyre (2016: 118) critiques United States legislation for its framing of doxxing as a stalking-related offence, as this places emphasis on what results *from* doxxing rather than validating the breach of private information as an offence itself. Across the legal jurisdictions of Australia and the United States, it is apparent that law does not sufficiently address the technology-facilitated component of doxxing and other forms of digital abuse. There is a clear need to counteract the 'legal lethargy' of current laws to address how harms are multiplied and compounded via the affordances of technology (Cinnamon, 2017: 609).

Australian and United States approaches to privacy rights are presently lagging behind the protection of personal data rights in Europe and the United Kingdom. The *General Data Protection Regulation* (GDPR) came into effect in 2016 in the European Union, and has set out applied protections for individual data rights in online and offline contexts (Hoofnagle et al, 2019: 66). The GDPR places the burden of good data stewardship, transparent collection and storage protocols, and equitable options to opt in and out of data collection on all institutions, organizations, and companies which collect data from users. Uniquely, the GDPR operationalizes privacy and data rights as an issue of human rights, where the 'inviolability of the person must be reconfirmed and reinforced in the electronic dimension, according to the new attention paid to the respect for the human body' (Rodotà, 2009: 81). The GDPR sets an important precedent for contemporary and future legal reforms in Australia and the United States, and highlights that a rigorous legal framework which supports privacy rights must centre the rights of the individual to full autonomy, transparency, and control over their personal data and information. While the GDPR is not a perfect legal instrument, and has been criticized for biases against working class and lower socioeconomic groups (Öhman, 2023), it sets a global benchmark for personal privacy protections in the present day.

Legal remedies for doxxing remain unclear and provide limited consideration for the experiences of victim–survivors. It is therefore useful to think about what law privileges and safeguards in its articulation of privacy. Sheila Brown (2006) offers a critique of the limitations of legal anthropocentrism. Identifying the need to move beyond the humanist parameters of law toward enriched and flexible legal possibilities, Brown (2006: 234) critiques that 'the body-in-law, so long a stable point of reference for interpreting and assessing the state of social order, has lost its definition, become blurred, unsteady, unsure, extensive'. Putting forward a case for 'granting equivalence to things and people', she calls for a richer legal framework to broaden the consideration of how embodiment in online space makes new order harms possible (Brown, 2006: 234). To engage with the full spectrum of people, things, and environments enhances the *imagination* of law: it bestows it with the vocabulary and ability to visualize the 'immanent and infinitely connected universe of quasi-objects' that comprise contemporary relations to the world (Brown, 2006: 233). An absence of this imagination leads to inadequate justice. In Chapter 3, a more detailed examination of the relationship between humans and objects will be outlined, in addition to a complication of the 'self' and the 'subject' beyond the unitary focus of the law. This foregrounds the importance of thinking about privacy as a *quality* of our informational selves, rather than a *condition* of how laws and legislation frame private lives, information, and entities.

Criminological and sociological framings of doxxing

In criminological and sociological scholarship, doxxing has been critiqued as a form of technology-facilitated abuse. Douglas (2016: 203–5) offered the first typology of doxxing, identifying three forms: *deanonymizing* doxxing, where the disclosure reveals a previously anonymized or pseudonymized identity without consent; *targeting* doxxing, where the victim is made vulnerable by information which reveals their physical location; and *delegitimizing* doxxing, which shames or humiliates the victim through the disclosure of information which transgresses 'established (or supposed) social norms'. In our scoping review, Wood and I identified a fourth type of doxxing – *disadvantaging* doxxing – which entails the loss of competitive advantage through the disclosure of proprietary information (Anderson and Wood, 2021: 210). These varied types of doxxing illustrate the vast range of information which can be captured and leveraged through non-consensual disclosure.

Digital criminologists have conceptualized doxxing as a form of technology-facilitated violence and abuse (TFVA), a framework which describes the use of technologies, digital devices, smart appliances, and social media to perpetrate harm and harassment. Doxxing has been discussed in conjunction with a variety of technology-facilitated harms and crimes, including technology-facilitated coercive control (Dragiewicz et al, 2018), image-based abuse (McGlynn and Rackley, 2017; Schriner and Rood, 2021; Schoenebeck, Lampe, and Triệu, 2023), stalking (Cohen-Almagor and Trottier, 2022), and misogynistic hate speech (Massanari, 2017; Eckert and Metzger-Riftkin, 2020; Huber, 2023). Digital criminology and technology-facilitated violence literatures have framed doxxing as an ancillary harm which is used to perpetrate other forms of cybercrime, such as cyberbullying, harassment, and image-based sexual abuse (Chen et al, 2019; Jørgensen and Demant, 2021). Bailey, Henry, and Flynn (2021: 2) include doxxing among examples of TFVA, which is 'carried out through text, images, and unwanted digitally-enabled or enhanced surveillance and monitoring, using a variety of devices and platforms from basic digital tools, such as texting, email, and social media, to more advanced technologies such as artificial intelligence'.

Digital spaces provide a seedbed for information disclosures which hinge on gendered relations between users, platforms, and networks. For example, reflecting on his research into far-right Discord groups, Jackson Wood (2024: 247) highlights how participants in these spaces utilize 'harmful normative discourses about race and gender' to facilitate their identity within the group, illustrating how the performance of harmful gendered attitudes can take root in normative uses of digital spaces. Lisa Sugiura (2021: 3) similarly critiques the normalization of 'anti-feminist rhetoric and ideology on platforms such as YouTube, Reddit and 4chan', cautioning that when these ideologies are 'presented as harmless or satirical media in the forms of images, videos and memes', gendered hate and misogyny take deeper root in digital spaces. This echoes what Emma Jane (2014, 2018: 662) identifies as the 'increasingly dominant online dialect' of abuse against women and girls, including misogynistic hate speech, e-bile and harassment. When normative cultures of information disclosure which denigrate women, girls, and gender-diverse people remain unchecked, platforms become complicit in facilitating gender-based digital violence.

While the discussion in this book primarily reflects on the experiences of the cisgendered men and women who participated in the interviews, it is vital to highlight that trans and gender diverse people experience unique forms of doxxing on the basis of their gender identity. I have recently written about the convergence of transphobic hate speech with doxxing attacks on popular social media sites like X,[11] which compounds the digital precarity of LGBTQ+ people (Anderson, 2025). Trans and gender-diverse people in particular experience unique identity erasure through doxxing when targeted through intimidating, misgendering and outing forms of hate speech (Anderson, 2025). In his research into 'digiqueer' experiences of harm, Justin Ellis (2023: 66) writes that doxxing campaigns against drag performers and trans allies highlights 'the pernicious and pervasive nature of doxing as a structured, technological harm'. The precarity and unique vulnerabilities that queer and trans people experience on the basis of their identities is more important than ever to highlight, particularly in the present political climate which pits trans and gender-diverse people against imagined binary-collectivism. Sarah Lamble (2026) describes this antagonism of trans lives as an example of 'carceral diagonalism', which sees a variety of political actors united to exclude trans and gender-diverse people from public life in an effort to protect a perceived 'vision of safety'. As anti-EDI rhetorics continue to take root in political and social life, it is vital that mobilizations against trans and gender-diverse people – including doxxing and privacy abuse – are identified, called out, and resisted.

Technology-facilitated violence and abuse research offer important criminological critiques and insights into digital harms; however, as Dunn (2021: 27) cautions, 'more precise terminology is useful when examining specific behaviours such as image-based sexual abuse, doxing, and digital stalking'. Focused and specific language to describe practices like doxxing is important as it situates the harm of doxxing as a valid form of abuse, rather than as a technique through which other forms of harm, like image-based sexual abuse or financial extortion using doxxed information, are performed. There is also a need to consider how LGBTQ+ victim–survivors 'experience the differential impacts of technologies within queer communities', resulting in unique and additional forms of harm through the leveraging of sensitive information about their sexual and gender identities (Fileborn and Ball,

2023: 10). While scholarship into LGBTQ+ experiences of technology-facilitated violence is growing in criminological and sociological research (see Farrell, 2021; Vogler et al, 2023), doxxing as a form of LGBTQ+ technology-facilitated abuse requires further engagement.

In addition to the harm of non-consensual disclosure, criminological and political science research has considered how doxxing might be used to hold powerful states, organizations and individuals to account. The nexus between doxxing and its potential to achieve justice through activism, whistleblowing, and political movements has also been evaluated by criminological and political science scholars (Garrick and Buck, 2022). Daniel Trottier (2019: 197) considers the 'pervasive and mobilizing' ways that doxxing can function as 'digital vigilantism' against oppressive power structures or corrupt public figures. While noting the justice potential of doxxing as a means of holding power to account, Trottier cautions that the same mobilization can lead to the reproduction of other harms such as misogynistic or racist hate speech, or other forms of cyberbullying.

Anne Cheung (2021: 577) similarly explores the justice potential of doxxing, noting that doxxing practices have increased in Hong Kong as a strategy of resisting and calling out police misconduct and corruption. Carmen Lee (2022) and Yao-Ti Li and Katherine Whitworth (2023) also discuss how political activists use doxxing as a strategy of resistance and refusal. In their framework of digital acts of resistance, Anita Lavorgna, Pamela Ugwudike and Yadira Sanchez-Benitez (2022: 265) expand Nielsen's (2003) concept of 'information pollution' to identify how the disclosure of 'malinformation' – that is, 'genuine information' which causes harm – can be leveraged by those with 'less power' against state and corporate entities. In other contexts, scholars have debated whether doxxing can be justified when used to out members of extremist and racist hate groups (Barry, 2020), with Kimberley Legocki, Kristen Walker, and Taylor Carriere (2018) suggesting that tensions arise between legitimate forms of citizen activism and grey areas of doxxing.

There are clearly numerous types and motivations for doxxing. Seven of the motivations for doxxing that have received the most attention from researchers include extortion, silencing, controlling, retribution, reputation-building, public interest, and unintentional doxxing (Anderson and Wood, 2021: 210–13). However, doxxing practices are dynamic and continue to evolve. While the motivations of perpetrators of doxxing reveal important insights into potential risks of non-consensual disclosure, ultimately it is essential that victim experiences are centred so that we might better understand how doxxing harms are experienced. There is little research which expressly considers how victims of privacy abuse, such as doxxing, conceptualize this harm and what they consider to be a just outcome in its aftermath. Exploring these perspectives is the key offering of this book.

This reflects our prior conceptualization of the imbricated nature of doxxing harms, where the riskiness of sensitive and personal information becomes actualized into ongoing, cyclical, and enduring threats to well-being and security (Anderson and Wood, 2022).

Highlighting how non-consensual disclosures overlap, compound, and imbricate over time is a novel method of mapping *how* the harms of doxxing are experienced. Doxxing is often viral in nature, meaning that the initial disclosure of information might go through an untold number of views, shares, reposts, and other communications. When thinking about the scale and context of a doxxing event, the virality of this form of communication confounds linear understandings of harm experiences having a beginning and an end. For victim–survivors of doxxing, they have no way of knowing where their personal, identifying, and sensitive information has spread, or who has access to it. Doxxing victim–survivors rarely achieve closure over the disclosure of their personal, identifying, and sensitive information. This, we term, is the second-order harm of doxxing: the risk of personal, identifying, and sensitive information being doxxed and re-doxxed, in perpetuity. We discuss first-order harms that represent immediate risks to the victim – this might take the form of humiliation, embarrassment, or pain that a doxxing attack provokes. Second-order harms, however, are located in the 'harm potentialities' of losing control over personal, identifying, and sensitive information in perpetuity (Anderson and Wood, 2022: 201).

(Post) digital harm? Understanding data abuse in changing and challenging technological times

Information is the currency of digital technologies, which have become indispensable to nearly all aspects of daily life. While innovations like the smartphone and the internet are not new, they have become embedded into the ways we perform our work, connect with family and friends, entertain ourselves, and garner insight into our bodies, moods, and habits. The rapid production of, and accessibility to, personal and identifying data has shifted the ways in which we understand ourselves and others. Wearable technologies like smartwatches, Bluetooth-enabled insulin monitors, and a raft of other technologically integrated devices have rendered our biometric data knowable, accessible, and more immediate than ever before (Jethani, 2015). Our integration with technology, however, does not solely provide benefits and efficiencies generated by data knowledge. As Shoshana Zuboff (2019: 9) critiques, the 'illusion … that being "connected" is somehow intrinsically pro-social, innately inclusive, or naturally tending toward the democratization of knowledge' has waned, and has been replaced with a growing concern for the harms that inhere within and through digital spaces. Digital technologies also prompt and create unique environments where distinct and complex harms can be perpetrated (see Elias, 2025). Framing information as inseparable from the self therefore sparks new questions about the role of technology in facilitating doxxing.

Concerns about the nature and extent of digital harms have arisen in academic, policy, and political contexts. From the 2000s and into the 2010s, scholars debated the degree to which digital harms might constitute 'real' risk and injury to victims. Peter Grabosky (2001: 243, 248) characterized online harms and cyber-enabled crimes as 'old wine in new bottles', arguing that while 'technologies may change rapidly … human nature does not'. This argument has since been contested by critical criminological, technology-facilitated violence, and feminist media scholars who note the severity, seriousness, and uniqueness of digital harm and abuse. Majid Yar (2005: 411), for example, identifies the 'force multiplier' of harms produced in cyberspace, where the affordances of technology extend the capacity of harm beyond a single and discrete incident to one that is at risk of going viral to large anonymous audiences. As Nicola Henry and Anastasia Powell (2015: 774) note, understanding the nature of digital and online harms requires that the 'techno-social lived experience' of digital harm be taken seriously so that 'harm can no longer be thought of only as related exclusively to the physical or biological body'. These framings align with what Luciano Floridi and colleagues (2015: 7) term the 'onlife', where our experiences of offline and online contexts become porous and blur into a hybridized experience of terrestrial and digital reality. As such, the validation of digital harms as 'real', significant, and worthy of further scholarly attention has shifted the focus of technology-facilitated violence research from binary framings of online and offline harm toward more expansive frameworks of harm mediation and technology-harm relations (Wood, 2025).

Digital harms are dynamic and ever-changing in response to the developments and innovations of technologies. Research into cyberbullying and harassment has also expanded rapidly in the last decade, with a focus on the capturing and mediation of image-based, video and audio sexual abuse, and, more recently, the harms of artificial intelligence (AI) generated sexual abuse (Ringrose, Regehr, and Milne, 2021; Pedersen et al, 2023). The impact of artificial intelligence and its capacity to produce simulated sexual images of victims without their knowledge or consent has shifted understandings of what constitutes types of harm like image-based sexual abuse (Flynn et al, 2022). These developments represent just some examples of the rapidly changing digital landscape in which harms arise and are inflicted upon digital subjects. Through these important research canons, debates about the nature, seriousness, and extent of digital harms have shifted in favour of victim-centred, affirming, and critical research.

While the severity of forms of digital harm like image-based, audio-based, and AI-generated sexual abuse has rightfully been recognized, doxxing in comparison has received less scholarly attention. In particular, there are crucial gaps in understanding how doxxing victim–survivors are harmed by the non-consensual disclosure of their personal information. This is not to say that there is no scholarship which has critiqued doxxing as a serious or noteworthy harm. For example, the work of David Douglas (2016), Åste Corbridge (2018), and Stine Eckert and Jade Metzger-Riftkin (2020) have all made important contributions and called for doxxing to be framed as a form of violence and technology-facilitated abuse. However, a gap remains in research which centres victim–survivor experiences and considers the harm of doxxing as something that lingers beyond the initial moment of disclosure.

Where previous studies have empirically researched types of doxxing, they have done so by quantitatively measuring 'rates of doxxing' in contexts like high schools and universities, with a particular focus on doxxing as a form of harassment or bullying between peers (Chen, Cheung, and Chan, 2019). Other studies have considered doxxing as a secondary harm clustered within other harmful experiences such as stalking through online dating apps (see Briandana, Oktavianingtyas, and Marta, 2020). Only a handful of studies have interviewed victims of doxxing with the purpose of understanding their perspectives, needs and justice desires. Eckert and Metzger-Riftkin's (2020: 284) research with victim–survivors of doxxing is the sole study which considers how the gender identity of victims intersects with misogynistic harassment of 'veillance cultures' online. Their research identifies a unique aspect of how personal, identifying, and sensitive information can be leveraged to harm along gendered lines. It is this lens that I pick up and extend throughout this book.

There is a clear need for further research into doxxing which centres the lived experience and perspectives of victim–survivors. I argue that doxxing exploits deeply personal, gendered, sensitive, and contextually complex information to impart harms that play out in similarly complex, unique, and nuanced ways. I build from the existing literature to explicate how victim–survivors conceptualize their experiences of being doxxed, how they feel harmed by these disclosures, and what they deem to be justice in the aftermath. Throughout this book, I grapple with how the harms of non-consensual disclosure are registered by victim–survivors as both an immediate harm and an enduring, cyclical risk to future well-being and security. By accounting for the non-linear, compounding, and intersecting nature of doxxing harms, I reveal new insights into what it means when autonomy over personal information is shattered, and the possibilities for resistance and rebuilding in the aftermath of privacy abuse.

The value of our information functions as a conduit to the free enjoyment and participation in digital spaces – a freedom which is ruptured by doxxing. In addition to understanding the harms of doxxing, it is vitally important to address the justice needs of victim–survivors. Presently, there are wholly unsatisfactory protections of personal information online and limited criminal justice responses when breaches occur. Understanding the harms of doxxing will also equip criminal justice agencies, governments, and policy makers with better strategies of supporting and responding to non-consensual disclosure. Engaging with victim–survivors on their own terms and through their self-definitions of harm generates new insights into doxxing scholarship and addresses the present gaps in research which values the perspectives of victim–survivors.

As our social lives continue to intertwine with digital technologies, it is vital that we understand the value of personal information and the ways that it is vulnerable to exploitation. This is particularly pressing, not only as a way of mitigating the risk of non-consensual disclosure in the digital contexts of our present but also in anticipation of the digital technologies and contexts yet to come. In addition to centring victim–survivor experiences, I engage with postphenomenological theory to understand how doxxing is performed and amplified through technologic design and functionality (see Wood et al, 2023). The readily available and accessible nature of data in the 'onlife' represents serious risks to the security of personal, identifying, and sensitive information (Floridi et al, 2015). Using postphenomenological theories of technology, I situate doxxing as harm that is embodied because it targets information that *constitutes* the ways that we inhabit, perform, and live as digital subjects. This departs from instrumentalist theories of technology which limit the influence of digital platforms, devices, and environments to that of a tool 'standing ready to serve the purposes of their users' (Feenberg, 2002: 5). In place of instrumentalist approaches to technology, I consider how technologies invite, afford, and extend capacities for harm online. Engaging an invitational approach to understanding digital harms accounts for the intersection of the agency of users alongside the capacity for harm that is embedded within the design of technologies (Wood et al, 2023: 1387).

Methodological approach and structure

This book engages with the testimonies and narratives of privacy abuse shared by 13 women and 5 men. The lived experiences of the 18 women and men who participated in the interviews map the terrain of how privacy abuse harms along gendered lines: through shaming and condemning women within a sexual hierarchy of worth; through the rupturing of control over personal information; while also sketching out the varied strategies of resistance possible. A brief glossary outlining the unique experiences of doxxing and relevant demographic information of the participants in the study is included in the front matter of this book. The findings shared by the 18 women and men arose from a qualitative, narrative-informed interview research project which sought to answer the following questions: how is doxxing experienced and conceptualized as a harm by victim–survivors? What are the motivations of doxxing perpetrators? Do legal, cultural, and ontological considerations of privacy and anonymity explain doxxing as a harmful experience? What are the justice needs of doxxing victim–survivors, and how can we best respond to them?

To answer these questions, I engaged a feminist, narrative-informed approach to interviewing. Adopting Cook and Walklate's (2019: 240) narrative method of interviewing victim–survivors, I prioritized conversational, tangential, and storytelling devices to '[shift] the focus from how moral transgressions are accomplished to how these moral transgressions are experienced'. Narrative approaches to interviewing focus 'not on the measurable nature and extent of violence but upon how violence is experienced and lived from the point of view of the victim' (Cook and Walklate, 2019: 241). In this study, the interviews were designed to engage the participants as 'active agents, exploring subjective meanings and experiences that have been ignored or silenced' or otherwise have gone unaddressed (Fraser and MacDougall, 2016: 244). Narrative techniques ranged from encouraging participants to speak about their experiences in their own words, to speak for as long as they wanted, and to push back or skip any questions that they did not want to answer. The research design and methods of this study were approved by the Human Research Ethics Committee at the University of Melbourne, Australia, in 2021.[12]

The book outlines the gendered, relational, and sensorial harms of doxxing in the following structure. In Chapter 2, I outline how doxxing is implicated in the digital social lives of victim–survivors. I highlight how more-than-human and information philosophy perspectives on data harms – and data *selves* – extend explanations of doxxing from merely being relegated to online, digital domains to being inseparable from the 'onlife' of the target. Bringing the insights of Rosi Braidotti and Luciano Floridi into conversation with each other, I consider the benefit of engaging with an *affirmative* ethical commitment to informational autonomy, as opposed to a deficit model of data harms. This theoretical and conceptual framework underpins the discussions of the experiences of the 18 women and men who had been doxxed, which are outlined in Chapters 3, 4, and 5. In Chapter 3, the violence of non-consensual disclosure is contextualized as a harm not only to data autonomy, but to the social and relational positioning of the victim–survivor. I explore how women experience targeted sexual shaming through doxxing, and develop a critique of the interpolation of 'shamed' victim–survivors at the bottom of the 'hierarchy of worth', a patriarchal value judgement about 'deserving' and 'unworthy' forms of victimization (see Vera-Gray, 2017). As such, the violence of non-consensual disclosure is inseparable from gendered harms which seek to denigrate women and children.

I continue this critical explication of the harms to informational autonomy posed by doxxing in Chapter 4, which sketches out the enduring, relational, and sensorial harms to privacy. Using the framework of *relational ruptures*, this chapter exposes the significant challenges to autonomous decision-making over personal information in the aftermath of doxxing. These ruptures also highlight the unique differences in experience between women and men, which forms the basis of discussion in Chapter 5. I consider the different access to support and solidarity that men who are doxxed reported, compared to the women in this study. However, I also note the complex ways that patriarchal norms of 'honour', 'strength', and 'overcoming' harm inhibit men's experiences of victimization.

In Chapter 6, I consider the agency of victim–survivors of doxxing to resist, withdraw, confront, create, and affirm their autonomy over their information. Strategies of digital disengagement, withdrawal, confrontation, and communities of support are examples of the techniques used by victim–survivors to resist and reclaim their informational autonomy after doxxing. The final chapter concludes this book by sketching out the theoretical framework of *informational autonomy*, which affirms the data rights of victim–survivors of doxxing. Doxxing is not the only risk to autonomy over personal, identifying, and sensitive information, but it sparks important reflections on the value that our information holds both now and in the future. Affirming the rights of all digital subjects to autonomy over their information is essential to revitalize privacy rights in the context of new and emerging technologies which exploit personal information. Affirming informational autonomy is also vitally important for ensuring that future generations have access to safe, respectful, and thriving digital spaces which do not leverage their information against their interests.

Conclusion

This chapter has addressed the legal, criminological, and sociological framings of doxxing to date. While doxxing is not the only way that privacy can be abused, it is important to foreground how privacy has developed as a legal, cultural, and social concept to understand parallel framings of doxxing in law. Understanding the development of privacy as a legal right also helps to highlight where current gaps and limitations exist in legal responses to doxxing. While laws can affirm and support victim–survivors of harm, I do not centre legal remedies to doxxing as a primary pathway to justice. I instead critique legal framings of doxxing and privacy as limited ways to understand what is valuable about personal, identifying, and sensitive information. The discussion of criminological and sociological research into doxxing points to important and developing understandings of doxxing as a form of technology-facilitated violence and abuse, as well as its potential for activism and informal justice-seeking by holding powerful states and actors to account. Building from the overview of understanding the harm of doxxing as an imbricated, intersectional, and enduring experience, the following chapter addresses what might be gained by thinking about privacy as a constitutive element of *informational selfhood*.

2

Personhood and Place in Online Domains

Introduction

In this chapter, I outline theoretical approaches to privacy which expand our understandings of the ways that technologies invite, afford, and sustain doxxing. Building from the critiques of legal framings of privacy in Chapter 1, I set out the key concepts and vocabularies which underscore this book's engagement with privacy abuse as a *posthuman* harm experience. First, I appraise Floridi's (2013) concepts of the ethics and philosophy of information, which sets out an *informational* framing of privacy. By framing subjects as informational selves, the ontic data of the subject – the data that 'constitutes' a person – puts into focus the risks of participating in onlife cultures (Floridi, 2013: 247). Drawing upon Floridi's ontology of informational selves, I argue that privacy is contingent on foregrounding respect, care, and reciprocity for the informational autonomy of all people.

With reference to posthuman and vibrant materialist insights from Braidotti (2019a, 2019b) and Bennett (2010), I situate the relationality between human, non-human, and more-than-human actors as a defining aspect of digital social life. Posthumanist and vibrantly material explorations of digital subjectivity contextualize relational and social harms, and foreground possibilities for resistance through an affirmative ethics of care and respect for personal, identifying, and sensitive information. I conclude this chapter by calling for a revitalization of anonymity as an individual right and foreground how anonymity is central to protecting the informational autonomy of all people.

The previous chapter reviewed legal and criminological understandings of privacy and its abuse through doxxing. These framings mark a valuable starting point in understanding the harm of non-consensual disclosure, but they do not give a full picture of how it is suffered, endured, and resisted against by victim–survivors. Legalistic, 'rights-based' accountings of privacy imagine a world where privacy is enjoyed equally and can be equated with property rights. Second-wave feminists have long been critical of how this ignores the gendered ways that privacy is used to exclude and control women, with Catherine MacKinnon (1987: 100) stating, 'there is no private, either normatively or empirically. Feminism confronts the fact that women have no privacy to lose or guarantee.' Taking up a 'rights-based' approach to privacy as something we all possess up until the moment of non-consensual disclosure leads to unfulfilling and incomplete legalistic responses – responses which, as Chapter 4 critiques in detail, leave victim–survivors of doxxing with scant options of seeking redress through the criminal justice system. A richer theoretical approach to privacy is therefore needed to make sense of the ways that technologies complicate and extend the experience of doxxing, particularly as a gendered expression of harm.

Building upon the critiques of legal framings of privacy in Chapter 1, this chapter foregrounds doxxing as a harm that is distinctly social. Chapter 2 begins with an overview of digital subjectivity, sketching out its interactions with privacy, anonymity, and human-technology relations. Drawing on vibrant materialist theories of embodiment and relationality, I play with and collapse the phenomenological understandings of 'place' in relation to digital space. This undertaking complicates the legal understandings of privacy as a personal right by accounting for the dynamic flows and relations between humans, objects, and technologies. By complicating the personal, this chapter critiques the reductive on/offline distinctions that treat embodied and mediated experiences as hierarchized and separate. In turn, this establishes a segue to the foundational theory frameworks of this book: Luciano Floridi's *philosophy of information* and Rosi Braidotti's *affirmative ethics.*

This chapter concludes by reflecting on Floridi's (2013: 212) framing of digital space as the defining characteristic of the 'information age', in which all subjects and objects communicate as informational entities, or 'inforgs'. Providing a functional vocabulary of Floridi's key terms, this section makes connections to posthumanist concepts of relationality, ethics, and action. In this section, I sketch out an affirmative model of informational autonomy, which foregrounds respect, reciprocity, and care between all informational actors. Reflecting on Braidotti's (2019a, 2019b) framework of affirmative ethics, I outline how the preservation and protection of the *contextual sensitivity* of information can promote relationships of 'ethical respect' and care in the aftermath of doxxing (Floridi, 2013: 259). This affirmative model of informational autonomy foregrounds the analyses and findings of Chapters 3, 4, and 5.

Digital subjectivity online

Reflecting on privacy requires a reflection on the public, both in the sense of *something* being public, as well as *someone* being a part of a public. Moving beyond the legal, proprietary, and personal framing of privacy, we can and should contest and reimagine the 'places' in which privacy and publics can be performed, and consequently breached. By playing with and collapsing the phenomenological understandings of 'place' in relation to digital space, I expand the notions of places in which digital subjects can both be public and constitute a public. Drawing on literatures of self-performance, digital social-spatiality, and vibrant materialities between humans and things, this section extends from the confining view of privacy as a personal and proprietary right. Instead, this section primarily reflects on the public assemblages of digital subjects. By first establishing how digital subjects constitute the public, this section explores how informational framings of privacy are key to understanding the ontological foundations of digital subjectivity. The informational ontology of privacy is further sketched out and conceptualized in the final sections of this chapter.

Performance of the self in online space is not a new concept, having received considerable scholarly attention (Chen, 2020; Peters, 2020; Freeman and Maloney, 2021). However, it is worth reflecting on who, or what, is considered to be a digital subject. Sarah Handyside and Jessica Ringrose (2017: 352) note that digital platforms provide tools to 'carefully [curate] ... idealized representations of users' lives', through which users edit, elevate, and, in some cases, mythologize elements of their lived histories. Warnings abound against falling for the 'idealization' of popular influencers online in literature ranging from the normalization of conservative gender norms (Sykes, 2024), the ethics and exploitation of family vlogging (Dunn, 2025), to education (Arceneaux and Dinu, 2018). A shared premise in these literatures is that the online persona is curated with *intention*: it is designed to be visible, read, seen, viewed, and engaged with by an imagined or known audience. The digital self is compiled with 'voluntary transparency' and is self-referential to the willingness of the subject to disclose aspects of their lived experience in 'deliberate acts of openness' (Lord, 2006: 17). However, the digital self is not only anchored by its purposeful construction but produces incidental and tacit data knowledge by being situated in online domains.

Online space is never neutral and is not composed of 'neutral bodies': it serves a neo-capitalist objective (Khoja-Moolji, 2015: 348; Zuboff, 2023). While the internet has sometimes been framed as an emancipatory utopia, Suneel Jethani (2015: 35) notes that this is always interred within the technocratic capitalism that orders, structures, and profits from online platforms. The tacit data knowledge collected by technologies – malware, bots, and algorithms, to name a few – enable 'outside actors to corral data into knowledge' about the digital subject, knowledge which is 'viewed through the lens of preexisting values' about marketability and profit margins (Heemsbergen, 2016: 140; Lord, 2006: 121). In this way, privacy is at the heart of both the structural and ideological function of digital space – the information of, and produced by, each subject is the product that makes digital spaces economically viable. However, digital subjects are not without options to resist against and subvert this process. Luke Heemsbergen (2016: 144) claims that 'radical disclosure' can be used to combat the pervasive power of capitalism and state power. In Heemsbergen's (2016: 144) account, 'voluntary disclosure of the secrets of others' is a device to destabilize the dominant political apparatus, stripping it of the power that secrecy yields. Here, visibility is invested as a strategy of seeking justice in response to the dominance of capitalism and data-extraction.

Although online space is underscored with capitalist values and objectives, it is also a space of great potential for the expansion of knowledge and connection between digital subjects. Digital freedoms are empowering, meeting 'the need for self-expression, the performance of identity, the chance to reinvent oneself, as well as the generosity of deliberate forgetfulness' (Floridi et al, 2015: 11). Gillian Youngs (2005: 69) terms the potential of online space as a 'sociospatial' elevation which transforms the way we 'do' work, how we socialize, and communicate with each other. As Jethani (2015: 35) outlines, the sociospatial possibilities of online space can operate as an 'emancipatory politics' that allows the digital subject to expand their embodiment by inhabiting a 'code/body'. This has unique implications for the generative social engagements that subjects can author and participate in, extending their abilities in postphenomenological ways previously limited to embodied performance. Peter-Paul Verbeek (2020: 142) highlights that human–technology relations have the capacity to generate new insights into power relations, political interaction, and the shaping of political mobilities through technology. Attending to these relations might '[open] the way to develop new ways of dealing with the social implications of technology in a democratic way' (Verbeek, 2020: 147).

While digital subjects can enjoy a raft of possibilities for the presentation and performance of the self online, we must be mindful of the 'compulsion' to participate: it is no longer voluntary but deemed essential to the management and maintenance of modern life (Kuntsman and Miyake, 2022). Treating online participation as a matter of 'compulsion' lends itself well to understanding the challenges and risks of digital subjecthood, particularly in the preservation of privacy (Heemsbergen, 2016: 138). A common solution to protecting privacy online might be to simply not post anything overly sensitive or personal. But this 'simple' fix does not account for the multiple ways that digital participation, disclosure, and transparency are required of our personal and professional identities. For example, LinkedIn networking or academic social media platforms can be used to build professional connections, seek job opportunities, or share research findings. Participation on these sites cannot exist without disclosure. It can be argued that the successful performance of professional identities, such as academia, public health, or other research work, necessitates a presence online. The digital self is not simply voluntary but can provide an essential extension of the individual's professional reach, performance, and capability (Stewart, 2016; Coad, 2017; Iliadis, Richards and Wood, 2020). The 'compulsion' to participate in online space cannot be dismissed as a superfluous engagement with the frivolities of social media but is instead underscored by the 'fullness' of self and subjecthood.

This raises several complex implications for theorizing and analysing the dynamics of digital subjecthood. First, the digital self is not easily dismissed as 'real' or 'unreal' but inhabits a far more complex *performance of selfhood* online, displacing the conditional attributes of an 'object' or thing possessed by the embodied subject. We cannot consider digital selves as 'objects' in the sense that digital mediums are intangible and constructed through sophisticated technological architecture. Instead, the digital self is an extension of the subject, remade in the image of our prior affective situation to others, but also in mirror relation to ourselves. Shenila Khoja-Moolji (2015: 348) reflects on this performance of self as an affective relationship through which selves are written into being in digital space by illustrating the 'impressions of history' that each user carries with them. Making use of these impressions, we engage in an affective reimagination in creating our digital selves as extensions of our lived histories.

A second complexity is the extent to which digital selves *matter* in the context of mediating and experiencing harms. Posthumanist scholars have long considered the generative potential of doing away with anthropocentric models which centre the human as the apex of knowledge, experience, and value. With my co-authors Kajsa Lundberg and Samantha O'Donnell, we conceptualize the messy entanglements of human, non-human, and more than human actors as a source of generative insight and a call for reflexivity in feminist research (Anderson, Lundberg, and O'Donnell, 2024). This framing of the generative potential of messy posthuman relations builds off Jane Bennett's (2010) theory of *vibrant materialism*, which treats all agents, human and non-human, as constitutive political entities. Subverting the actor–object hierarchy, Bennett (2010: 2, emphasis in original) highlights 'the active role of *nonhuman* materials in public life', terming this reciprocation of cause and effect between human, non-human, and more-than-human actors as 'thing-power'. To use Bennett's terms, digital subjects, as material selves, *do* matter – they have 'thing-power' as entities that can both enact and receive communications, information, and action. This conception of the 'thing-side of affect' is not new, but draws upon a rich philosophical vocabulary, such as Latour's notion of the ecology of human and non-human elements as the *collective*, and Spinoza's account of the *conatus*, or possibility for power, within living things (Spinoza, 1992; Latour, 2009). As is discussed in the final sections of this chapter, the relations between human, non-human, and more-than-human actants have much to offer our engagements with and understandings of digitally mediated harms like doxxing.

The 'onlife': informational anonymity as a revitalization of privacy

Thus far, this chapter has refuted legal interpretations of privacy, complicated the iterations of self and subjecthood, and considered the affective possibilities of relationships between humans and things. Just as Bennett approaches her task of examining the vibrant materialism of things as 'lavishing attention on [them]' (2010: ix), I attend to the complications of publics, personhood, and place through Luciano Floridi's *philosophy of information*. Floridi et al's (2015: 1) philosophy illustrates the shifting parameters of privacy in the 'informational age', a period of postmodernity where the on/off line distinctions collapse to merge together as the 'onlife'. This section introduces some key definitions and vocabularies, foregrounding the ongoing engagement with Floridi in the methodological design of the project and its analysis of data.

Floridi's philosophy of information considers the emergence of *the informational age*. This age marks the 'fourth revolution of self-understanding', following the Copernican, Darwinian, and Freudian revolutions, where humans are reinterpreted as 'informational organisms living and interacting with other informational agents in the infosphere' (Floridi et al, 2015: 54). The infosphere is akin to the 'pre-industrial village', where all citizens were known and knowable to their neighbours. Importantly, Floridi (2013: 240) distinguishes between the infosphere and the village as the 'digital has the means to protect what the small village must necessarily forfeit'. Similarly to Bennett's vibrant materialism of things and objects, Floridi (2013: 212) considers that all actors, human or otherwise, are *inforgs* – 'informational multi-agents' – capable of containing, transmitting, and receiving information in dialogue with each other. In the informational age, we cast 'digital [shadows]', where the 'hybridization' of our lived and mediated realities have coalesced into entirely new experiences of subjecthood (Floridi et al, 2015: 45).

Subjecthood is mediated in what Floridi et al (2015: 8) calls the *onlife*: the 'hyperconnected' imbrication of reality, whether it be in lived or virtual realms, which produces new experiences of human connection, engagement, productivity, and communication. The onlife problematizes the binary distinctions of online spaces being 'real' or 'virtual'. As Floridi et al (2015: 91) point out, these distinctions invoke Cartesian dualism, where the non-cognitive, 'extended' body is 'divorced' from the 'non-extended … fully disembodied mind'. This book is also couched in a shift from the Cartesian dualism model – whereby the 'body' comes to represent lived physical experiences and the 'mind' reflects the disembodied online life. This framing of the onlife helps us better understand how embodied, intangible, and digital experiences of selfhood intersect and co-exist simultaneously.

Floridi's framework provides new conceptual insights into privacy as an issue of data sensitivity and access. By framing subjects as informational selves, the *ontic data* of the subject – the data that 'constitutes' a person – gives insight into the inherent tensions of participating in onlife cultures (Floridi, 2013: 247). Floridi terms this tension as the *ontological friction* between agents who oppose the informational flow within the space of information. To Floridi, ontological friction is a necessary tension, as it produces the agent's ability to limit or enable the flow of their information to other agents in the system. This tension is foundational in establishing Floridi's (2013: 235) claim that informational privacy is a function of the ontological friction in the infosphere, as it frames the realms of private and public as contingent on the *choice* of actors to relay their information with others. A functional definition of privacy in the informational age is as follows: 'Privacy is the right of individuals … to control the lifecycle … of their information and determine for themselves when, how, and to what extent their information is processed by others' (Floridi, 2013: 236).

This framing of privacy is especially insightful to the research aims of this book. I add to this definition, by considering that the revitalization of anonymity as an individual right is chiefly important in the protection and promotion of informational autonomy. While being online has been held to mean being anonymous, this is no longer accurate: a raft of malware, spyware, and cookies have transformed the screen from 'a shield for privacy … [to] a window onto lives online' (Floridi, 2013: 239). Clearly, the challenges posed by the informational age extend the current conceptual thresholds of privacy and anonymity beyond their utility. In the next section, I briefly outline the importance of revitalizing anonymity as a key aspect of the affirmative ethics of care, respect, and reciprocity required to foster informational autonomy, building upon Floridi's framing of informational selves.

Earlier, I noted that the neo-capitalist and post-democratic realities of online space and surveillance challenge ideas of privacy as a proprietary and personal right. To know is to contain, which greatly furthers the objectives of the capitalist project to control its consumers. If we consider that the onlife complicates the distinction between public and private domains, we must confront the realities of surveillance and control in all aspects of life in the informational age. Surveillance studies have extensively documented the systems of control that govern our movement in cities (Lippert, 2009), our online purchasing habits (Goldfeder et al, 2017) and even the intersection of our biometric data with location technologies (Jethani, 2015; Lupton, 2016). Further, the tensions of the informational age promote the surveillance of the self through bio-health interfaces, such as heart rate monitoring, kilojoule intake, and sleep patterns (Wolf, 2016; Lupton, 2019). In the onlife, divides between offline places and digital spaces converge, making digital subjects vulnerable to surveillance wherever they are.

Processes of informational surveillance fit within Helen Nissenbaum's concept of the 'socio-technical' ordering of informational life. Nissenbaum finds three approximate purposes for technology in the information age: (1) to monitor and track, (2) to analyse and aggregate data, and (3) to disseminate and publish information (Nissenbaum, 2009: 20). When constant surveillance is enacted upon you and also *by* you, the thresholds of anonymity seem inaccessible and tenuous. If we take anonymity to be the process of being 'forgotten', this seems at odds with the architectural affordances of digital space, and indeed, the flow of information between agents in the information age. However, an information-oriented anonymity framework is not simply concerned with restricting the flow of information but 'ensuring that it flows appropriately' (Nissenbaum, 2009: 2). Returning to Floridi's tension of ontological friction between agents, anonymity must be revitalized to account for the *contextual sensitivity* of information between agents, a sensitivity that deserves care and respect.

So far, this section has established that the informational nature of our world and our selves present unique implications for the framing of privacy and anonymity. Drawing upon Floridi's ontological friction, we can extend conceptualizations of privacy from the proprietary ownership of an individual right to the necessary tension of communication abilities between agents. In the context of these communicative abilities, anonymity has the potential to be actualized as a productive *and* protective force, as it reflects the contextual sensitivity of information. Nissenbaum (2009: 37) notes that 'information begets information' in a flow of production, as it 'yields implications, consequences and prediction' from the aggregation of data read in context. If we take information as a productive and generative force, then the need for anonymity as the *choice* of an agent to not participate in this flow of information is more important than ever. Floridi (2013: 242) points out that 'one's personal information is not lost when acquired by someone else', but it is *actualized* against the agent's ability to participate or withdraw from the ontological friction between agents. Here, the acquisition of personal identifiable information is harmful because it enables the further perpetration of identification, by unknown agents, in perpetuity.

Concessions to personal, identifying, and sensitive information are dangerous because they are difficult to reclaim. While the agent has not 'lost' their personal information when it is shared with others, they lose their ability to control who has access to it – an echo of Floridi's (2013: 236) 'right to control the lifecycle of data'. Floridi (2013: 247) also claims that 'arbitrary data' about a person, such as their social security number, PIN, or name, is not ontologically linked; it is 'merely associated with someone's identity and can easily be detached from it without affecting the individual'. While this is true, arbitrary data are still features of personal identity, and therefore risky. The connection of ontic data to 'arbitrary' data must be noted, too – the HIV status of an anonymized patient in a database is inconsequential, but if the patient is de-anonymized, connected to the patient ID, and breached by a user of the database, then the facets of identity combine to constitute a harm. By framing anonymity as the 'relation of caring, not by the mere amount or type of information exchanged', a conceptual reimagining of 'rights' and a re-ontologizing of subjects in the onlife is stimulated (Floridi, 2013: 254). This illustrates that anonymity must be affirmed as a right of individual choice, to accommodate not only the agent's choice to be visible but the relations of care that protect the contextual sensitivity of information.

Contextual integrity

This section will engage with Nissenbaum's concept of contextual integrity in further detail and will explore its potential for an affirmative framework of privacy as a form of informational autonomy. Nissenbaum (2009: 130, 2004: 137) considers that our social communication and engagement takes place in *contexts*, or 'structured social settings with characteristics that have evolved over time', and that within these contexts, social life is governed by 'norms of information flow'. Informational norms include standards of professional and personal address, cultural communication, and modalities of interpersonal communications, to name a few. Importantly, 'informational norms regulate the flow of information of certain types about an information subject from one actor (acting in a particular capacity, or role) to another or others (acting in a particular capacity or role)' (Nissenbaum, 2009: 141). Information 'flows' through processes of disclosure and distribution, which concerns the 'movement or transfer of information from one party to another'. Through the movement of information through communication, norms of distribution build upon the importance of the appropriateness of information disclosure by ascertaining whether its 'distribution, or flow, respects contextual norms of information flow' (Nissenbaum, 2004: 141). Nissenbaum (2009: 140) introduces the qualifier of 'integrity' by supposing that in an idealized situation, contextual integrity is 'preserved when informational norms are respected and violated when informational norms are breached'. As such, she infers that integrity is the starting point, which becomes sullied and assailed when the norms of informational flows are not respected or properly exercised.

What is an improper informational flow? The contextual integrity of privacy relies on two types of informational norms – the 'norms of appropriateness and norms of flow and distribution' (Nissenbaum, 2004: 138). Norms of appropriateness concern how information disclosure is contextually dependent and may shift in degrees of appropriateness from one setting to another. Nissenbaum uses the example of disclosing medical details to a doctor in the clinical context of a medical facility to illustrate the appropriateness of that information, as opposed to the inappropriateness of being forced to share private medical history with an employer. While norms of appropriateness address the multiple transformations that information might have in its various contextual disclosures, Nissenbaum notes that appropriateness is not a characteristic of the information but of the setting in which it is disclosed. As such, 'appropriating information from one situation and inserting it in another can constitute a violation', such as participating in a Mardi Gras celebration in your private life but not being openly out in your workplace (Nissenbaum, 2004: 140). The disclosure of this personal information from a person who bridges both contextual environments would constitute a privacy breach. Similarly, norms of distribution require that information is not shared to parties beyond the purview of the individual's consent. In the scenario of disclosing medical details, we might consider that the non-consensual disclosure of sensitive medical information to an outside party, such as the patient's sexual health, reproductive history, or HIV status, would be a breach in the norms of distribution and would therefore constitute a rupture in the flow of information.

The merits of applying contextual integrity to privacy abuse are compelling, although there are many aspects of this model that remain unexplored. For example, the contextual integrity rights of individuals might struggle to be considered within the practice of meta-data scraping, where innocuous data including scroll-time, engagement with advertising materials, and hyperlink click-through data is sold for market research (Zuboff, 2022). Further, the consequences of remedying contextual integrity rights with national security legislation that warrants surveillance of its constituents might also be explored in further detail. Presently, contextual integrity has featured in computational science research, particularly in projects that discuss behaviour in online social networks (Criado and Such, 2015) and the challenges in bioethics and big data (de Groot, 2024). Contextual integrity has also featured in research about the privacy concerns of Internet of Things household smart appliances (Apthorpe et al, 2018), the ethics of using big data for research and development purposes (Zimmer, 2018), and in examining the 'permissions' function on new-generation smart phones (Wijesekera et al, 2015). Eckert and Metzger-Riftkin (2020) have made some important connections between contextual integrity and the privacy harm of doxxing, but a harm-informed framework of privacy as a form of contextual and informational integrity needs further expansion. The final section of this chapter considers this question, returning to posthumanist framings of the need for an affirmative ethics in doing so.

Posthumanism, care, and affirmative ethics

Earlier in this chapter, I discussed framings of relationality between human, non-human, and more-than-human actors. These framings offer important theoretical understandings of digital life, subjectivity, and experiences of technology, particularly through critiques and possibilities of the use of social media and other digital technologies. In the final section of this chapter, I consider how posthumanism – the 'epistemological framework for supporting the elaboration of alternative values and new codes of inter-relation that extend beyond human influence and cognition, but do not discount it' – contextualizes relational and social harms, and foregrounds possibilities for resistance through affirmative ethics of care and respect (Bignall and Braidotti, 2019: 2). This affirmative ethics inspires a key offering of this book: how an affirmative model of informational autonomy might respond to the harms of non-consensual disclosure and offers strategies for resistance, justice seeking, and recognition.

Rosi Braidotti, a prominent theorist and scholar of posthumanism, has pioneered affirmative ethics as a posthuman ontology, modality, and pedagogy. Braidotti's (2019a: 141) 'affirmative politics and ethical relations' between human, non-human, and more-than-human actants calls for a 'differential relational ethics of embodied differences'. In Braidotti's (2019a: 141) framework, a differential relational ethics is fundamentally posthuman because it does away with 'the notion of undifferentiated unity, totality, and One-ness', acknowledging that the ethical needs of human, non-human and more-than-human actors will undoubtedly be diverse. Elsewhere, Braidotti (2019b: 470) frames affirmative relations and ethics as an emancipatory process of affirming 'our innermost freedom – the freedom to become all we are capable of, all our bodies can take'. In contrast to the limiting and negative instantiation of rights through law which frames harm through what it *takes* from the subject, Braidotti's affirmative ethics calls for us to imagine what our freedom from harm can and *should* be imagined to be.

Engaging with affirmative ethics of care for information is vital in a climate where distinctions between bodies, selves, and digital lives are porous and shifting. Jannice Käll (2017: 1151) has noted that the 'dematerialization' of bodies as corporeal stands to '[reshape] both what we perceive as humans as well as what we perceive as data'. The implications for affirmative, liberatory ethics and politics toward our information is therefore both expansive and timely in meeting these challenges. Affirmative ethics not only offers a vocabulary through which we might critique power dynamics, violence, and abuse, but it also articulates strategies of recovery, resistance, and autonomy in the aftermath of experiencing harm. By adopting Braidotti's framing of the affirmative potential of 'freedom', we might understand our information as key to the ways that we conceptualize, protect, and celebrate privacy as an ontology of digital subjectivity. Braidotti's model of an affirmative politics intersects with Floridi's call for 'relations of care' for our information and the information of others. Like Braidotti, Floridi (2013: 132) describes this care as a collective approach, describing the infosphere as 'the fastest-growing environment that human beings as informational entities are going to share with other non-biological informational entities'. In calling for a robust model of information ethics, Floridi (2013: 132) argues that 'new generations are going to need a mature sense of ethical responsibility for and stewardship of their whole environment, both natural and artificial (that is, in one word, informational), to foster responsible care of it rather than despoliation or mere exploitation'.

Responsible care of information calls in the actions, respect, and care of all actors within the informational environment. Floridi (2017: 391–2) considers this care to be emerging alongside theorizations of 'ethical infrastructure' or 'infra-ethics', where the design of informational computer technologies are not neutral but imbued with the values of their designers. Elsewhere, my co-authors and I (Wood et al, 2023) similarly reflect that the design of digital technologies invite both intended and unintended uses which can result in harm or violence. While some abuses of technology leverage their affordances to harm in ways that were unintended by the designer (see Wood, 2025), the complicity of technology and its design in making forms of harm possible must also be considered. For example, Pavan and Lavorgna (2021: 548) highlight that the 'diffusion' of image-based abuse material is a '"more than human" form of violence, perpetrated in and through "human-non-human assemblages"' that are made possible by technologic and relational networks of connection. Responsibility in the information age, therefore, is bestowed not only upon the individual actor but conceptualized as a broader, porous, and relational interaction between human actors, technologies, systems, and other more-than-human entities.

An affirmative model of respect for and protection of information calls for revitalized, individualized, and reciprocal protection of information in digital environments. This could include abolishing metadata extraction on popular social media sites, often hidden in baroque terms and conditions of use statements. Respect and reciprocity for information might also look like transparent practices when data is scraped and sold, including royalty or residual models where users are compensated for their data (Hacker, 2018). Open-data activists have long praised the benefits of embracing an open access model of data sharing, claiming that greater tranches of data are imperative to the expansion of machine learning and the development of artificial intelligence (AI) (Tegmark, 2018). In the era of generative AI, posthumanist and informational philosophers are also querying the ways that we might narrow and refine our definitions for what 'counts' as data, and how we can empower individuals to safeguard and protect 'their' personal information (Käll, 2024: 189). These potential models of reciprocity and respect are, for the time being, speculative in nature. What this chapter develops is the potential for reciprocity and care of information as an affirmative model of resistance in the aftermath of doxxing.

Braidotti (2019b: 477) considers the confederation of human, non-human, and more-than-human actors as 'affirmative transversal assemblages'. It is through these assemblages, she argues, that we can 'extract margins of possible action that would enable us to trace what we are in the process of becoming … and construct patterns of alternative becoming' (Braidotti, 2019b: 477). Calls to action and to patterns of alternative becoming, I argue, can be situated within relations of care, respect, and stewardship over personal, identifying, and sensitive information. The liberatory potential of *informational autonomy* is one such way that victim–survivors of doxxing can embody resistance in the aftermath of non-consensual disclosure. What Braidotti and Floridi offer to this practice is a language to understand how relations of reciprocity, care, and custodianship underscore this autonomy. Autonomy over information must be foregrounded in respect for the information of *all* actors. I set out this model of an affirmative model of privacy rights, understood through respect for contextual determination, autonomy, and choice of the individual actor, to build from the important work of Braidotti and Floridi, and to develop an exploration of what we might put forward as informational autonomy in the aftermath of doxxing harm.

Conclusion

The insights provided by dismantling proprietary–personal privacy rights are essential as 'we fear and reject what we fail to semantacise' (Floridi et al, 2015: 3). This illustrates the need to reconceptualize privacy rights and protections in the information age. Inflexibility in understanding how privacy abuse harms its victims is the cause of ineffective legal action and justice responses. In turn, this justifies a radical reframing of rights, moving toward an affirmative ethics of information as a model of care and intimacy between all actors, human and more-than-human. To stimulate this revitalization, we must ask radical questions: what information is worth protecting? What information is capable of being protected? How must we reimagine the self to properly articulate the rights of the modern, informational age subject? Building on the framing of intimacy as 'care', understandings of digital social life must pivot to reflect that informational entities deserve respect and consideration. After all, when dealing with 'personal and private information, [we] are dealing with human beings themselves' (Floridi, 2013: 259).

This chapter has established that privacy in the information age requires the revitalization of anonymity as an individual right and calls for an affirmative model of informational autonomy. While the need for privacy has not been disputed, in this chapter I have argued that legal approaches to understanding the harms of doxxing are insufficient. Moving towards a new ontology of informational selves provides the necessary complexity for engaging with the reconceptualization of privacy as an informational right, and affirmative autonomy over information as the means to protect it.

3

The Violence of Non-Consensual Disclosure

Introduction

'Knowledge is power', or so the adage goes. What we might take this to mean in the context of doxxing is that to use knowledge of someone without their consent and against their interests is to hold power *over* them. Holding 'power over' another entails 'the ability to control others, events, resources, or one's self in spite of resistance—rather than alternatives such as the ability to cooperate, to give freely of oneself, or to feel and act in harmony with nature' (Johnson, 2005: 39). In confronting the power embedded within disclosure, this chapter outlines how the non-consensual disclosure of doxxing relays harm through the imposition of power *over* its victim. Earlier research has framed doxxing as a conduit for interpersonal violence, facilitating harms like image-based sexual abuse, coercive control, harassment, and cyberbullying (McGlynn and Rackley, 2017; Dragiewicz et al, 2018; Huber, 2023). However, engagement with non-consensual disclosure as a *form* of violence and harm warrants further attention.

This chapter argues that doxxing produces violent ruptures to social life among families, partners, friends, and colleagues. Social ruptures deny the doxxed individual autonomy in their decision-making about how they participate in social life, with whom their sensitive information is shared, and in what social contexts it is made accessible. Ruptures to informational autonomy are traumatic events, scrambling and mixing up the individual's prior framework of reference for disclosure, requiring a new mitigation of risk of 'unbottled' information, which must be negotiated amidst feelings of intrusion and insecurity. Similar to a 'rupture', Brison (2017: 224) reflects on the nature of trauma as an event that 'not only shatters one's fundamental assumptions about the world and one's safety in it, but also severs sustaining connections between the self and the rest of humanity. Victims of human-inflicted trauma are reduced to mere objects by their tormenters: their subjectivity is rendered useless and viewed as worthless.' This chapter will develop a conceptual model of the violent rupture of doxxing not only as a technique of harm and intrusion, but a challenge to the *perception of self* of the individual. This represents the enduring and overlapping violence of doxxing: as a harm event, a degradation of autonomy, and a threat to the perception of self of the victim–survivor.

Drawing on 18 interviews with victim–survivors of doxxing, this chapter considers the gendered ruptures of doxxing are key in understanding it as a form of violence. Differences in types, experiences, and outcomes of doxxing reproduce a 'hierarchy of worth' among doxxing victims, a practice which highlights both the explicit and 'mundane' types of misogyny that have become 'socially sanctioned and normalized' (Tranchese and Sugiura, 2021: 2711). While men may have different access to support from family and friends, this chapter suggests that women are interpolated within different timelines and consequences of doxxing, as indicated by barriers to justice-seeking, validation from family, and grappling with their own self-blame for their experience.

Informational harms, violent ruptures

The violence of doxxing ruptures autonomy over information, ceasing the ability to freely participate in social life in the contexts and manner of the subject's choosing. In its place, a different form of relationality – of undermined agency, unmitigated future risk to second-order doxxing, and fear and anxiety – reorders the relational ontology of the subject to their social bonds. The aftermath of harm has been described as a 'shattering' of the self (Brison, 2002; 2017), where 'relational ruptures' occur due to the 'failed witness' of finding recognition and validation from community after experiencing trauma or abuse (Benjamin, 2022: 52). In some cases, this rupturing might be abrupt, where doxxed information explodes into virality, leading to large-scale shaming and harassment. In other cases, the rupture of doxxing might be slow – a domino falling into place behind a long line of pre-existing social alienation. Nixon's (2011: 2) concept of 'slow violence' – a form of relational violence in which 'delayed destruction … is dispersed across time and space, an attritional violence that is typically not viewed as violence at all' – is useful to consider how the temporal rhythms of doxxing rupture victim–survivors' social bonds in the present, as well as in the future.

Of the 13 women interviewed in this study, 11 had experienced a form of image-based sexual abuse, disclosure of their reproductive history, or other form of gendered shaming through doxxing. Image-based sexual abuse (IBSA), which involves the threat, creation, and dissemination of sexual, private and sensitive photographs, videos, and audio, has received critical feminist scholarly attention in recent years (McGlynn and Rackley, 2017; Yar and Drew, 2019; DeKeseredy, 2021; Hayman, 2024). The intersecting harms of image-based abuse, bullying, and harassment have elsewhere been framed as acts of cyber-located sexual violence (Hayman, 2024), particularly in technology-facilitated violence and abuse literatures (Dragiewicz et al, 2018; Powell et al, 2020). Image-based doxxing has been increasingly receiving attention as a risk factor for women and children in digital spaces, particularly as concerns mount over the complicity of technologic platforms in 'shaping an environment that can be more or less conducive of online IBSA' (Pavan and Lavorgna, 2021: 556). With generative artificial intelligence (AI) becoming increasingly sophisticated, the scope and scale of image-based doxxing now no longer requires access to a pre-existing personal or sensitive image or video. Simulations of individuals in sexually explicit content can now be generated with an otherwise ubiquitous photograph of a target's face or profile, which are distorted through 'nudification apps' or other technologies which use generative AI to simulate the target in a sexual scenario or form of abuse material (see di Silvestro and Iurissevich, 2024). As technologies continue to develop, the risks of image-based doxxing will also continue to shift and change.

In the following section, the experiences of Odette, Aisha, Tegan, and Tamara are outlined. These accounts highlight how, for many women, privacy abuse coincides with the non-consensual disclosure of their sensitive images and videos, as well as other information which makes them vulnerable to further forms of targeted harassment and vilification. Not only do these findings point to highly gendered *types* of doxxing, where the information disclosed without consent was highly sensitive to the women's bodily autonomy, sexual lives, and medical histories, they also illustrate the gendered *violence* of social rupture. Social ruptures create rifts and divides in family life, friendships, workplaces, and other social contexts. Doxxing violently disrupts familial, interpersonal, and intimate social bonds in both real time and futures to come. Ruptures also reveal and exacerbate previously existing hierarchies of worth in these relationships – hierarchies which are mired in patriarchal framings of shame and blame.

Many victim–survivors of image-based doxxing report that the disclosure of their sensitive photos and videos exposed them to judgement, harassment, and shame from their communities. For Odette, her doxxing led to prolonged sexual shaming and condemnation from her family and friends. Odette experienced image-based sexual abuse via doxxing when a video of her dancing at a nightclub 'partly naked' was filmed without her knowledge and uploaded to Instagram and Facebook. Odette described the impact of going viral as a prolonged form of social shaming:

> Everybody who shared it used it to give advice to their children. Used it to give advice to their wives, their sisters. So I became a laughing stock, for a lesson being learned. (Odette)

Odette's doxxing brought about many consequences, including losing her job and falling out with her family and friends. Notably, the shame she experienced underscored many aspects of these consequences. Vera-Gray (2017: 129) describes the pitting of women against each other as a product of the 'hierarchy of worth' which '[situates] women and girls in relation to each other, as well as in relation to men and boys'. These hierarchies 'interact and intersect with gender inequality', differing 'according to other markers of a woman's or a girl's social location' (Vera-Gray, 2017: 129). Gendered hierarchies of worth have become deeply entrenched in what Collins and Rothe (2023: 311) describe as the 'endemic misogynist *normal*' (emphasis in original text), reflecting social and political debates about women's rights to bodily autonomy and the increased popularity of extremist incel and far-right ideologies (Sugiura, 2021: 76). For Odette, she became situated at the 'base' of the 'social location' of acceptable constructions of femininity, pitted against an idealized child, wife, and sister as the 'cautionary tale' of what happens to women who rebel against the social norms of modesty and conservative sexuality.

Social constructions of shame around female sexuality and self-expression often prevent the recognition of image-based doxxing as a form of abuse. The rhetoric of 'responsibility' is integral to the victim-blaming of targets of image-based doxxing, which positions them as 'ruined' and sullied (see Uldbjerg, 2021a). This rhetoric has been rife in high-profile image-based doxxing cases, such as the iCloud leaks of personal photos of celebrities like Jennifer Lawrence and Emma Watson in the early 2010s (Suzor, Seignior, and Singleton, 2017). More recently, the (mis)use of generative AI nudification tools has led to celebrities, streamers, and lay-users being targeted with simulations of abuse material (Gibson et al, 2024). As technologies become increasingly sophisticated, and 'responsibility' rhetorics grow murkier in the era of AI, there are further opportunities to understand how sexual shaming motivates and contextualizes image-based doxxing.

It is increasingly important to be critical of the digital spaces and locations which enable and permit the abuse of personal data through doxxing. 'Internet intermediaries' – that is, service providers, platforms, and social media conglomerates – are complicit in sustaining locations of violence where digital harms, like doxxing, are not only possible but *commercially lucrative*. Digital spaces are locations where women and gender-diverse people are not only vulnerable to gender-based digital violence (Suzor et al, 2019) but are spaces where their abuse is profitable through increased user engagement, encouraged by lax platform governance (Gjika, 2023: 7). Reflecting on the barriers to removing abuse content online, Pavan (2017: 67) writes that:

> In not tuning their [terms of service] in a way that allows them to either tackle or sanction such occurrences as actual forms of abuse,

> intermediaries are in fact exerting a hidden influence in shaping an online space of interaction that is not only tolerant towards but, more relevantly, reinforces discriminatory and abusive behaviours against women.

Of the 13 women who shared their experiences, nine reported that their doxxed information was disclosed on an internet intermediary like WhatsApp, Instagram, and Facebook. This highlights that social media platforms are not only deeply entangled with the sharing of abuse material, such as doxxing, but also reveals how technologies are both socially shaped and socially shaping in moments of digital abuse (Williams, 1974). The following section will explore how shame is socially produced and mediated in the aftermath of doxxing, leading to the violent rupture of relationships and communities of support for the victim–survivor.

Blame, shame, and payback

Intense feelings of shame and blame often follow doxxing for women who experience intrusions of their personal, identifying or sensitive information. For many of the women in this study, blame and shame in the aftermath of doxxing was experienced not only externally from their communities but as an internal self-judgement. Many participants noted that they were not only ashamed that their intimate images and videos were exposed but that the resulting ruptures in the form of judgements and harassment brought about additional and serious forms of shaming. For victim–survivors like Odette, the overwhelming shaming she endured made it difficult for her to recognize her experience as one of victimization. Instead, she placed the blame upon herself. Odette lives in Australia, where image-based sexual abuse is a criminal offence. When asked if she considered reporting the sharing of the video to the police, Odette shared that:

> I just didn't want to escalate more shame. I was already feeling ashamed. I was broken because I lost my job. I didn't want to escalate things. I didn't have the emotional strength for everything. I blamed myself for everything. So I saw no need of going to the police. (Odette)

For many victim–survivors, the perception that they had nowhere to turn to for support characterized the challenges of moving forward. Eila, who experienced image-based doxxing after a 'disagreement' with her partner, reflected that 'when family judges, friends judge … who will listen to you, who can you rely on? No one.' These forms of shame are not only experienced at an interpersonal level, but also impede the ease of participating fully in their communities and cultures. These ruptures to social life not only harm through the experience of feeling blamed or shamed for their doxxed information but take away possible communities of support.

The social ruptures of doxxing also harm by highlighting gendered stereotypes and standards within the victim–survivors' social world. For Aisha, intimate photos doxxed by her ex-partner represented a serious risk to her reputation within her culturally and religiously conservative community. Aisha was doxxed by her ex-partner in the aftermath of their break-up, an experience that she defines as a form of payback when she chose to leave the relationship. 'Retributive' forms of image-based doxxing function as an 'identity-consolidation practice', which are justified by the perpetrator as a means of reifying their power over the victim–survivor in the aftermath of a perceived denial (for example, here sparked by the break-up) (Henry et al, 2021: 97). The backlash Aisha experienced when her religious community saw her doxxed intimate photos was intense, and she experienced forms of harassment and judgement. In Aisha's words:

> A lot of people said I wasn't supposed to have those pictures and I wasn't supposed to have things like that on my phone or give them out to people, no matter the kind of relationship I was having with them. I felt like I made a really big mistake and it was such a sad experience for me and I felt really disappointed in myself. (Aisha)

The public censure of Aisha by her community echoes Romania's (2022: 353) description of image-based sexual abuse as a '[vector] of shame … mainly used as a source of gossip, a form of social control that joins the private and public spheres'. Aisha's doxxing disrupted not only her personal and private relationships but also ruptured her public standing within her community. Vera-Gray's (2017) conceptualization of the 'hierarchy of worth' is again relevant here, with public sentiments toward victims of image-based sexual abuse denigrating them to the bottom of the hierarchy as a consequence of rejecting 'normative' codes of modesty and the denial of female sexual agency. The sexual lives of women have historically been framed within the privacy (and control) of domestic spaces, with feminist reconstructions of gender roles subverting scripts about female desire and sexual agency (Ackelsberg and Shanley, 2018). As such, ruptures resulting from image-based doxxing violently upend women like Aisha's sense of community and sense of self.

Many religious and conservative communities impugn the freedom of women's sexual and self-expression. However, as Ghumkhor (2019: 120) reminds us, fantasies of 'the whole' woman often spur racist and jingoistic imaginings of forms of abuse experienced 'over there' by non-western women. I take this framing in mind here in an effort to avoid making generalizations and assumptions about the power implicit within Aisha's experience as a woman in a conservative religious society, especially in her accounting being denied sexual agency. Werbner (2007: 165) also gives an explanation of how 'notions of honour and shame are located at the point where familial politics and the politics of religion, tribe, and nation meet. Their significance is thus inevitably ambiguous, dynamic, and shifting.'

The intersection of religious, family, and cultural values is evident in how Aisha described her shame after the retributive doxxing attack. Throughout the interview, she indicated that she had retreated from public life, which included deleting her social media accounts and not reaching out to friends for fear of being judged. Aisha expressed that she was 'lucky that her whole family didn't get to find out', with only her mother being a source of support during this time. Aisha's fear of judgement from her friends and family reflects how the proliferation of stereotypes about 'good' and 'bad' women pose further harm to victims of image-based doxxing in religious or culturally conservative communities. No longer coded as 'the asexual, obedient woman, viewed through the lens of ideologically class gendered notions of appropriate behaviour', Aisha's experience of being doxxed simultaneously harmed her control over her personal and intimate information while also exposing her to serious and ongoing consequences to her social standing within her community (Hussain, 1997: 202). This illustrates how the social rupturing of doxxing plays out in both intersubjective and interpersonal contexts.

Other women also reported that their doxxing was perpetrated as a form of payback for perceived social transgressions. In Tegan's interview, she shared that a close friend disclosed information about her abortion to her parents by making a public post on Facebook and 'tagging' them:

> I approached her when I saw it online and when my parents talked to me about it. But then she was just so proud of what she was doing. She didn't see anything wrong with that. According to her, she said she wanted to pay me back. That I always have most of the things she wanted, but I didn't know that. And I don't know what she wanted that I took from her. I don't know. (Tegan)

Shaming rhetorics about reproductive health, particularly abortion care, are often rooted in conservative and religious value frameworks but are also mired in stigma about the 'type of woman' who has an abortion (Røseth et al, 2024: 84). Instead of drawing upon a political explanation of 'conservative' or 'progressive' people seeking abortion care, Kumar et al (2009: 4) suggest that female sexuality is 'at the core of abortion stigma because it may amplify transgressions of stated norms about who, when, why and how to have sex'. Regardless of the subtext of values in Tegan's case, her account of how doxxing harmed her hinges on a patriarchal judgement about her sexual life. Tegan noted that her family, particularly her father, were 'very disappointed' in her decision to terminate the pregnancy and did not speak to her for many months. Tegan's experience of doxxing is therefore intertwined with several forms of harm – the harm of losing informational autonomy about a private medical decision, the harm of experiencing shame, judgement and distancing from her family, and the harm of knowing that a previous friend was responsible for her abortion becoming public knowledge, and was 'proud' of it.

In Tamara's case, her perpetrator was also a close friend, who sold highly sensitive intimate images of her, in addition to her home address and contact information.

> I tried to confront her. At first she was lying. She lied and lied and lied. Then she made a statement like 'Everything is always all about me.' What does that have to do with my picture? What does that have to do with my home address? What does that have to do with my number? Why share everything in my life to strange people? (Tamara)

Tamara described the images of her that were doxxed as 'coarse' and 'very sexual, with over 90% of my body open'. It was not clear during the interview whether Tamara took these photos herself or whether they were taken without her consent. Regardless, the disclosure of this information intersected with other forms of harm. Not only was she betrayed by a close friend perpetrating this form of doxxing, but she also endured the intense feeling of shame and degradation after having highly sensitive intimate images shared online and the additional harm of having this information 'sold'. Throughout the interview, Tamara indicated that she was very concerned about her safety and well-being after being doxxed:

> I keep getting strange calls at that moment. I mean, everybody kept letting me know they have my number. They know where I live. I wasn't feeling safe. I wasn't feeling secured. I mean, there were times where I went to a few friends' houses to sleep. I wasn't safe, I wasn't secure. I felt like my family members were also in danger. I mean, anything could happen. (Tamara)

The combined disclosure of Tamara's intimate images, personal contact information, and home address created an enduring state of panic, suspicion, and unrest that extended well beyond the initial moment of doxxing. This reflects what Wood and I term as the 'second-order harms' of doxxing, where the target of doxxing experiences setbacks to their 'objective and subjective security interests' (Anderson and Wood, 2022: 201). In Tamara's case, she had genuine concerns for her well-being and physical security, as she was receiving sustained and ongoing harassment via repetitive phone calls and people 'turning up' at her house. This can be classified as an objective first-order harm to her security interests, as her well-being and physical security at home were compromised. Additionally, Tamara reported a subjective second-order harm, where her ability to feel safe and trust that she was secure was undermined. Tamara's reflection that 'anything could happen' demonstrates the subjective second-order harm of anxiety – where the 'harm potentiality' agitated by experiencing doxxing is 'actualized simply through learning that one's PII has been compromised' (Anderson and Wood, 2022: 201). Crucially, Tamara's experience of the second-order harm of anxiety was not predicated on physical injury *actually* befalling her or her family. Instead, the erosion of trust, the feeling of losing control over information, and the resulting anxiety of who might use that information for unknown purposes sketches out the ongoing and enduring harms that victims of doxxing experience.

Erosion and erasure of self-identity

So far, the ways that doxxing violently ruptures reputation and status within familial or social networks has been mapped out. These are serious forms of harm which directly impact the social freedoms of victim–survivors of doxxing. In addition to rupturing the individual's ability to exercise autonomy over the contexts, locations, and people to whom they disclose sensitive information, doxxing is dually harmful because it ruptures the individual's *perception of the self.* Informational abuse, elsewhere termed in domestic and family violence scholarship as 'identity abuse' (see Donovan, Magić, and West, 2024: 1342), has been identified as a form of abuse where the victims '[have] their identity disregarded or used against them'. Insights from domestic and family violence literatures are vital in drawing together an analysis of how the non-consensual disclosure of personal information hinges upon patriarchal value judgements of worth and 'respectability'. This final section will reflect further on Aisha, Tamara, and Odette's experiences of internalizing the consequences, shame, and reputational harm following their doxxing as an experience of *self-blame,* requiring new modes of 'undoing' and 'remaking' the self in the aftermath of trauma (Brison, 2017: 224).

For victim–survivors like Odette, experiencing the disclosure of deeply personal sensitive information changed how her outlook both outwardly and inwardly.

> [Doxxing] has done nothing but produce severe harm to me, it has brought a lot of pain, pain that I would never forget in my entire life. It's something that has made me sick and almost cost my life. So I just feel like it has a way of, it has a way of causing people much pain, distress, and a lot more. It has done nothing positive to me in a painful

> way and has given me a very hard hearted mind. Nothing would ever look or sound painful or hurtful, more than being doxxed. (Odette)

Odette's reflections highlight how digital harming lingers long after the event itself may have 'concluded'. Elias (2025: 46) argues that the impacts of digital social harm require an attentiveness to 'the overt ways in which the digital context mutates the modes and mechanisms of power'. Odette's profound and lingering pain in the aftermath of her doxxing draws focus to the significance of personal data abuse. Her ability to connect with her immediate, known social networks was ruptured, denying her support and care from her community. This rupture also prohibited her connection with other potential digital networks, leaving her with 'a hard hearted mind'.

Aisha's account also illustrates how the imposition of power took on new digital implications of harm. After her ex-partner doxxed her intimate photos and videos in a retributive form of image-based doxxing, she reflected on the difficulty of assuaging her own guilt and sense of shame:

> At the time, I felt really devastated and I was depressed for a long time because I kept on blaming myself and thinking I was stupid for letting someone have my videos and photos. So I felt depressed at the time. (Aisha)

In addition to wrestling with her own internalized judgement of blame – for sharing intimate images and videos with her partner in a previously consensual manner – Aisha also highlighted how many women who experience digital sexual violence also struggle with identity in the aftermath of their experience. These challenges of struggling to fit within categories of 'victim' or 'survivor' are additionally complex when the individual is harbouring an internalized judgement of their worth (or in this case, the perception of being complicit in the 'creation' of their abuse material) (O'Shea et al, 2024). In the following excerpt, Aisha details the internalized self-judgements – of being 'too good' – which hindered her from seeking support from her friends:

> I chose not to talk about it [with friends]. Cause I felt like they would obviously say I wasn't supposed to and I was too good to have. They were going to judge me, so I just felt like I shouldn't talk about it. (Aisha)

Like many victim–survivors who have endured sexual shaming or harming, Aisha's difficulty in reconciling her perception of herself as 'good' was brought into tension with her judgement that those who experience image-based doxxing do not often garner empathy (Adair and Senn,

2025). These self-judgements have been described by Evteeva, Burges, and Gelabert (2024: 82) as a form of internalized misogyny, where women co-opt patriarchal value judgements against themselves and other women in a form of 'psychological oppression'. Insights like those from Odette and Aisha highlight that experiences of image-based doxxing change how victim–survivors must renegotiate their self-concept in the aftermath of harm which lingers long after the initial event.

Other victim–survivors, like Tamara, outline how stigma 'sticks' to victim–survivors, even when they do not experience internalized self-blame and shame:

> Both now and then looking into the future, it-it's just had and keeps having a deadly impact on me. A negative impact on me. It's more like a dirty stigma. It's a crime that will stick on me for the rest of my life. I could go anywhere and I'll be presented my nude and my contact details that were shared everywhere. (Tamara)

The 'stickiness' of stigma that Tamara reflects on here sheds light on the challenges for victim–survivors to join up their past selves with their present and future reality post-doxxing. 'Stickiness', a conceptual framework first sketched out by Ahmed (2014: 87) in thinking about the 'histories of contact between bodies, objects and signs', might give a lens to view how informational harms linger long after first contact. In their research into feminist activism in digital spaces, O'Neill and Fileborn (2025: 12) note that the performance of disgust online 'becomes 'sticky' and attached' to men who harass women and girls online through repeated 'performative utterances' of disgust. This is useful not only in understanding how doxxing might be an example of a performative utterance and a launching pad or accelerant to secondary harassment stemming from the disclosure. Additionally, when stigma 'adheres' to victim–survivors like Tamara, we gain insight into how the stickiness of shame comprises both the perception of shame from the eye of the beholder, as well as the individual. Feelings of dirtiness, of imposition – negative impacts – take away from Tamara's unimpeded self-expression – or just 'being' – in digital and offline social spaces.

The 'stickiness' of shame in the aftermath of doxxing can also play out as a sensation of feeling 'stuck' in the moment of harm. For Tamara, the lack of resolution post-doxxing placed her in a similar double-bind, where her desire for 'vengeance' could not be realized. This led to a feeling of being 'stuck' in place, tethered to the moment of harm with limited opportunities to move on.

> I, I felt like my hands were tied. I felt like I couldn't just do much for myself. I felt like I wasn't there for myself. I felt like, I mean, we all

> know that when you are being hurt by someone, most times if your spirit and your soul seeks vengeance, until you go, until you avenge whatever other person did to you, most times you'll not find peace. So I didn't even find peace with myself. (Tamara)

Without resolution like seeing her perpetrator brought to justice, Tamara's anger became trapped with few avenues for release. While Audre Lorde (1997 [1981]: 280) identified that anger 'can become a powerful source of energy serving progress and change', Kay and Banet-Weiser (2019: 606) highlight that for many, pathways to progress and change are elusive. They write that when 'expressing anger [has] not worked; modulating anger [has] not worked', the inability to find resolution results in an 'affective injustice … [which] has us trapped in a double-bind'. Tamara's unresolved efforts to 'find peace with [herself]' show how the 'double-bind' of being unable to fully 'express' or 'modulate' anger or pain in the aftermath of doxxing further erode the self-concept of the individual.

The loss of informational autonomy over personal information also represents a form of self-erasure. Future possibilities of consenting, sharing, and determining the contexts in which personal information is accessible are eroded by the rupture of doxxing. This also limits the expression of the self – in relationships, family dynamics, in community – in a manner which becomes contained *by* the non-consensual disclosure. Tamara reflected on the enduring challenges of navigating her family relationships after her intimate photos were doxxed:

> Well, my relationship with my family, I kind of feel like it has come back to normal, but not that normal. Cause anytime I wrong anyone in my family, they tend to bring up that issue and it gets me down again. (Tamara)

Tamara highlights how the impact of being doxxed has altered the ways in which she engages with her family and loved ones, in a dynamic that gives the appearance of being 'normal' only to be taken away in the instance of conflict. These are the enduring challenges that doxxing victim–survivors must navigate – the possibilities of their presence and performance of relationships becomes filtered through the lens of how their personal, sensitive information has recontextualized their status and standing in their social lives. Tamara described her life being 'shattered and [broken] apart' by her doxxing attack. This echoes what Ahmed (2016: 22) describes as the erosion of attacks to identity, where the self becomes 'chipped away at' by personal attacks that '[hammer] away at our being'.

Doxxing that erodes and erases the personal identity of the target is not hyperbolic but increasingly apparent in the context of political mobilization against diverse, equality, and inclusion measures in the United States and beyond, influenced by Trump's administration. This positions trans women, trans men and gender diverse people to be particularly precarious to informational abuse which erodes at their sense of self and autonomy. I have elsewhere (Anderson, 2025) reflected on the harms of erasure doxxing, highlighting that:

> disclosures which seek to destroy-through-naming (e.g. shattering privacy through pronoun harms) are more than words that name – they represent affirmations, performances, realities of trans lives. Disclosure, therefore, is a 'double valence' of destruction because it dually shatters the autonomy and control over personal data while purporting to deny the target the autonomy to make themselves known in the manner, contexts, and to the people of their choosing. (Anderson, 2025: 13)

While the experiences of Tamara, Tegan, Aisha, and Odette reflect the harms of misogyny toward cisgendered women, it is important to highlight that these struggles are interconnected with those of 'transphobia, femmephobia and trans-misogyny' (Colliver, 2021: 231). Engaging with the violence of doxxing, therefore, requires an attunement to the complex ways that patriarchal, misogynistic, and hateful speech acts harm all women – cisgendered, trans, gender-diverse and gender-queer – in interrelated, yet distinct ways.

Conclusion

This chapter has set out some of the intersections that doxxing shares with violent and harming acts. Whether doxxing coincides with sexually violating disclosures of information, retributive dissemination of personal medical histories, or efforts to discredit the reputation of the target, the impacts of doxxing are experienced through gendered constructions of worth, shame, and blame. As the accounts of Tamara, Odette, Aisha, and Tegan have revealed in this chapter, doxxing cannot be understood without reference to the imposition of power through acts of non-consensual disclosure. 'Power over' the women through the disclosure of their PII occurred in ways that specifically leveraged the precarity of their information in gendered forms, and to gendered affects.

As Tegan's experience highlights, women experience an additional precarity to their informational autonomy due to the gendered vulnerabilities of health and reproductive data. This is particularly heightened in a political and social climate which is regressing the medical and privacy rights of people who seek abortion care and other reproductive freedoms. Presently under the Trump administration, the right to seek abortion care was repealed, and police have been granted new powers to access personal devices to access personal medical information through period tracking apps to 'prove' that a termination occurred (Nelson, 2024). Similarly, in the United Kingdom, the National Police Chiefs' Council (NPCC) have amended their recommendations of review following miscarriages after 24 weeks to allow them to seize personal devices and access personal medical records (McCudden, 2025). Non-consensual disclosures of personal reproductive data represent a specifically gendered form of data violence which can result in more than informational intrusions, including the criminalization of healthcare seeking in regressive, authoritarian contexts. The sensitivity of gendered, medical information also spotlights the role of technologies, like period-tracking apps or 'diaries', in potentially becoming vectors for data intrusions and abuse (Conti-Cook, 2020).

The use of doxxing to leverage sensitive sexual information must also be critiqued as an explicit form of violence, and effort to hold 'power over' women and gender-diverse people (Johnson, 2005). As the experiences of Aisha, Odette, and Tamara illustrate, the retributive disclosure of sexual images and videos to 'pay back', 'punish', or 'take down a peg' all centre on motivations to diminish the victim–survivor. It is worth noting that while victim–survivors like Aisha report that her perpetrator was an ex-partner motivated to harm her after their break up, both Odette and Tamara were doxxed by previously close female friends. 'Power over' framings often draw upon patriarchal conceptions of leveraging power to dominate women (see Ortner, 2022), but accounts of sensitive information doxxing shed light on the complicity of women in perpetrating harm against other women. By highlighting how 'internalized sexism causes women to compete against one another rather than banding together', motivations for retributive doxxing are expanded beyond a male-perpetrator/female-victim dichotomy (Rosida et al, 2022: 14). Internalized misogyny occurs not only when women blame themselves for experiencing harm and abuse, but also offers an important framework for understanding the motivations of female perpetrators of image-based doxxing and other gendered forms of privacy abuse.

This chapter has also outlined that doxxing is not just a conduit to experiencing other forms of cyber violence but a unique harm experience which violates autonomy over information, sense of self, and security and freedom in online spaces. For women like Aisha, Tegan and Odette, the internalized blame and shame, as well as the externalized judgements they faced from their communities, prevented them from seeking help and support. These social ruptures not only deny victim–survivors of doxxing vital access to care from their communities after experiencing doxxing but sustain an ontological 'wound' to the self, requiring them to 'remake' themselves (Brison, 2017). Understanding doxxing as a unique form of violence, therefore, is integral to building a critique of privacy harms as non-linear, which is developed further in Chapter 4. By understanding doxxing as an ongoing and cyclical form of violence, non-consensual disclosure shapes 'what we are ceasing to be' as informational subjects, in addition to 'what we are in the process of becoming' in the aftermath of disclosure (Braidotti, 2019c: 36). Understanding privacy harms as a form of violence is not hyperbolic but foregrounds an important language to recognize how personal data, and the decision-making we exercise over our information, underscores how we express, present, and mobilize ourselves online. Violent ruptures to this form of identity are significant and require recognition.

Privacy is not merely a 'negative' right – the right to 'prevent' knowledge from entering the public domain, or to prevent the 'constellation of intrusions' in contemporary surveillance but a 'positive' right because it provides for the free, autonomous explication of the self (Nissenbaum, 2020: 31). Many privacy theorists and philosophers include autonomy as a defining rationale for the protection, mandating, or safeguarding of personal privacy rights (Westin, 1968; Gavison, 1980). Privacy over personal, sensitive, and identifying information is utterly integral to understandings of gender-based violence because it underscores that acts of abuse and harm are about holding dominion and power over another. To frame doxxing as a harm to informational autonomy, therefore, is a powerful means of sketching out the potential for an affirmative ethics of care for information. As the following chapter details, an affirmative ethics of care also captures the complex ways that data harms require response and recognition both in the present, and futures to come.

4

Privacy Harms in the Present and Future

Introduction

Doxxing sparks a large range of violent relational ruptures which flow on from the initial moment of disclosure. In this chapter, I map out a continuum of privacy harm which traces the disruptive effects of data intrusions to both the present and future autonomy of the target. Moments of doxxing rupture the *possibility* of future informational autonomy, where the future 'potentialities' of non-consensual disclosure become more risky and likely to be actualized (see Anderson and Wood, 2022: 197). These ruptures to informational autonomy are significant in their own right but take on additional co-morbidities of risk when doxxing is deployed to harm in a gendered manner. Gender-based doxxing, as I have termed it previously in Chapter 3, ruptures informational autonomy for women and gender-diverse people in unique and targeted forms. The reflections of Tamara, Carol, Odette, Jana, Molly, Hara, Aisha, and Jessie discussed in this chapter not only highlight the gendered *impacts* of shame, loss, and self-blame but also underscore that women and gender-diverse people are targeted by uniquely gendered *types* of doxxing.

Attending to the gendered harms of doxxing, its impacts, and its shaping of violent social ruptures does not hierarchize any type of experience over another. However, it is clear that victim–survivors are differentially positioned within a patriarchal hierarchy of value judgements which seeks to position to stratify victim experiences against value judgements of 'goodness' or 'worthiness'. There are myriad ways that patriarchy underscores social relationships, and gender-based doxxing further entrenches these hierarchies of worth and systems of dominance. Patriarchy harms all people – women, gender diverse and queer people, and children, as well as men who do not conform to the exacting standards of masculinity required by patriarchal systems of control (Idriss, 2021). This chapter aims to examine the varied and nuanced ways that these harms are experienced, not in an effort to compare or belittle different experiences but to draw out the different iterations of patriarchy in moments of social rupture, and to consider how ruptures to informational autonomy are motivated by efforts to dominate and diminish the target.

Building on Chapter 3's analysis of the gendered hierarchy of worth, in Chapter 4 I consider how doxxing poses risk and harm to the target in the future, as well as the present. The accounts of the eight women outlined in this chapter trace how sexual shaming, ostracization from family and friends, and fall-out in their professional and working lives further entrenched hierarchies of worth where they were judged against patriarchal values. This was particularly complex for women who had intersectional religious and cultural backgrounds, which added an additional lens of condemnation, shame, and self-blame. Family support, or its absence, was an instrumental factor in how the women characterized the consequences of being doxxed. Nearly all of the women experienced some kind of familial alienation (Scharp and Hall, 2017), isolation or shunning (Lewin, 2000), or conditional support from close family members. For the women, being doxxed through digital sexual abuse and other disclosures which sexually shamed them positioned them as undesirable and therefore not worthy of support (Uldbjerg, 2021b). This chapter situates gender as enmeshed within the social bonds of family identity, motherhood, sisterhood, intimate friendships, and romantic relationships.

Ruptures as social judgements

Women who experienced doxxing through image-based abuse reported tacit and explicit moments of judgement from their family, friends, and community. Tacit moments of judgement might take the form of a 'pedagogy of regret', where victims internalize judgements about their 'sullied digital footprint', feeling that the 'inevitable public shame' they are subjected to is deserved (Albury, 2017: 714). Responses to tacit judgement varied across the participants, with some, like Aisha, choosing to withdraw almost entirely from their social lives, whereas others felt like they were losing an 'impossible battle' of having people in their lives take their stories seriously. Tamara reported the following when describing the difficulty of being heard:

> Everybody judged me without hearing my own side of the story. People looked down on me dirty, people looked at me like I was crazy. People looked at me like I had no life to live anymore. People looked at me like I should go die, like I should go commit suicide. (Tamara)

The intense feelings of judgement experienced by Tamara manifested in serious ways for her mental health and well-being. While Tamara did not personally identify that she felt suicidal in the aftermath of her doxxing, she was judged by her community as a 'coarse' and 'sullied' woman with nothing to 'live for'. Additionally, Tamara's perception of being judged was intensified due to her belief that 'everyone' had seen her intimate photos. This reflects what Uldbjerg (2020: 110) terms the spectre of 'the potential onlooker', where victims of digital sexual assault cannot determine precisely who has seen the intimate image of them, meaning 'that anyone they meet on and off digital media can potentially be the holder of a shaming look' (Uldbjerg, 2021b: 4). This 'shaming look', characterized by Tamara as perceiving her as 'dirty', 'crazy', and deserving 'death' is yet another example of how gendered constructions of power underscore women's experiences of doxxing victimization, shaping the ways that ruptures ripple across their social lives. As noted by Powell and Henry (2014: 123), image-based abuse harms need to be more clearly situated as an issue of 'consent: where it is the behaviour of those who violate an individual's sexual autonomy (including over their sexual or intimate image) that is shamed' and not the victim. Shame and blame, therefore, become operationalized in highly gendered ways in the rupturing of victim–survivors' social bonds, rather than perpetrators.

Of the 11 women who reported being doxxed through the sharing of intimate photos and videos, five of them knew who their perpetrator was, with three being male ex-partners and two being close female friends. This finding is significant as much of the doxxing literature considers anonymous perpetration to be a defining feature of 'the power imbalance between perpetrator and victims' (Chen et al, 2019: 219). By considering instances of doxxing where the perpetrator *is* known to the victim–survivor, this finding further supports the argument that doxxing is relational and contingent upon the social norms, gendered values, and cultural understandings of worth which underpin relationality under patriarchy. Some perpetrators were reportedly 'proud' of their actions, while others explicitly called their doxxing a form of payback for perceived social wrong doing. While image-based sexual abuse research has primarily focused on the harms suffered by victims, emerging scholarship into the framings of responsibility by perpetrators indicate that a variety of intentions can underscore moments of image-based abuse (Powell et al, 2022).

In the case of Tamara, where a close friend sold intimate photos of her as a form of payback, the perpetrator might be defined as 'being unable to control their need for revenge' or 'needing acknowledgement' from their peers (Harder and Hasinoff, 2021: 9). The use of doxxing to pay back Tamara for her perceived higher social status reflects a desire of the perpetrator to substitute the praise that Tamara received with judgement that the perpetrator thought she deserved. This could function as a form of 'acknowledgement' for the perpetrator, who gets to play out their fantasy of 'shedding light' on Tamara by denigrating her vis-à-vis doxxing. Operationalized through payback, doxxing here ruptures social bonds by denigrating Tamara to a lower status in the hierarchy of worth. This models what Schippers (2007: 94) terms 'hegemonic femininity', where norm production about what a woman *should be* '[legitimatizes] a hierarchical and complementary relationship to hegemonic masculinity … [guaranteeing] the dominant position of men and the subordination of women'. Here, we can observe the complicity of women within patriarchal systems of domination, where hegemonic femininity is operationalized to justify the rupturing of social life through doxxing as a means of upholding the patriarchal status quo.

Sexual shaming in the aftermath of doxxed intimate images conveys serious harm to victim–survivors of image-based doxxing. The women reported intense judgement and social isolation from their family, friends, and communities. The challenge of feeling that they had 'let down' or 'embarrassed' their families mapped onto the belief that they had done something wrong by taking intimate images or videos in the first place. The distress produced by these ruptures was particularly impactful to the women's status and standing in their family lives, and resulted in serious mental health consequences, including depression, anxiety, not knowing who they could trust, and suicidal ideation. It is important to distinguish that the shame experienced by the participants – or indeed, the shame placed upon them by those in their social spheres – does not reflect that they have done or 'are' something shameful. As Vangelisti (2006: xv) notes, the 'staying power' of family and community can be both a resource and a restriction to those who experience harm and abuse. While the 'staying power' of relational bonds can 'increase [family] members' commitment to each other and the family as a whole', the 'involuntary nature of family relationships' can also pose a risk to victim–survivors of abuse who may become enmeshed in unsupportive or unforgiving family dynamics which amplify their experiences of alienation and rejection (Vangelisti, 2006: xv). Nikki Rutter (2023) has also reflected on the 'ruptures' that can emerge in moments of adolescent child-to-parent harm, highlighting the multi-directionality of familial relational structures, and the lasting impact that these bonds can have in shaping experiences of abuse.

Paying attention to the underlying influence of patriarchy is one such way to resist these learned processes of self-shaming and blaming. Reflecting on what Uldbjerg (2021a: 28) terms the 'stigma of the "ruined" victim', we need to prioritize framings of victim experiences that do not sideline them as 'inherently damaged by the assault and thus deprived of agency in creating better futures for themselves'. Reflecting on the dynamics of sexual shame provides an important grasp on how gendered harm and power take root in the ruptures of doxxing. The intersecting harms of being doxxed anonymously must also not be ignored. Uldbjerg's (2020: 110) 'potential onlooker' is relevant here to illustrate the twofold harm of not knowing who perpetrated a doxxing attack, nor why they were motivated to do so. Many participants felt that their perpetrator wanted to undermine or diminish their reputation, particularly in a professional context. Carol speculated that her doxxing, which involved the leaking of confidential company information that she was responsible for, was predicated on diminishing her professional reputation and standing within her workplace:

> I don't think the person wanted to blackmail me. Why I'm saying that is because if this person wanted to blackmail me, this person would've reached out to me first. Would've tried to reach out to me first, maybe to ask for some money or to ask for some impossible things. But nobody reached out to me. So I think this person was trying to undermine me. I see myself as a very principled person. I feel this person was trying to undermine me. He has a lot to do with undermining me, humiliating me, and embarrassing me. Or even getting me sacked from work. (Carol)

As Carol notes that she is a 'principled' person; doxxing confidential company documents misrepresented her competency and professionalism. Here, the harm of doxxing undermined her skills and placed her at risk of professional consequences. Carol also sets out an interesting distinction between using doxxing to 'blackmail' and using it to 'undermine' – she notes that because the perpetrator did not reach out after her doxxing occurred to request money or some other demand, that their motivations must instead have been to undermine her professionally. Some doxxing scholars, such as Douglas (2016: 202), do not consider that doxxing can be used as a form of blackmail. However, Carol's account here brings complexity to these distinctions, as her doxxing is not without financial consequences. Indeed, Carol suggests that she was at risk of losing her job as a consequence of being doxxed. The motivation for undermining her professional reputation intersects not only with financial implications for her livelihood but with the social dynamics of reputation. While Carol does not refer to reputation directly, she implicitly touches on the impacts of doxxing to her reputation by noting that she felt 'humiliated' and 'embarrassed'. To experience humiliation and embarrassment points to a general desire to present yourself on your own terms. Here, we can observe how diminished informational autonomy draws together myriad financial, interpersonal, social, and integrity harms. Importantly, Carol's reflection sketches out how the second-order harm of being doxxed also intersects with other digitally enabled harms like blackmail.

Doxxing motivations include a range of targeting, silencing, retributive and malicious intentions – and when deployed in a viral or coordinated campaign, can erode the possibilities of informational freedom (Anderson and Wood, 2021; Anderson, 2025). As Odette reflects, her doxxing was a sophisticated and planned attack involving her being filmed without her knowledge or consent:

> It made me feel as if I was being monitored. I think this whole thing didn't just start at that moment. It had started a long time ago. (Odette)

This reflection gives an interesting insight into how the anonymity of a doxxing perpetrator adds an additional dimension to the harm experience. The suspicion of being monitored for a 'long time' in the lead-up to the disclosure added additional insult to injury in the aftermath of being doxxed. In Odette's case, the non-consensual recording was released on social media, in addition to her home address and workplace information. This resulted in Odette losing her job in the aftermath of experiencing intense social shaming. For Odette, knowing that her doxxing was premeditated – likely to be a person close to her who knew her home address and workplace – amplified her feeling of distress. This maps onto what Boehnlein et al (2020: 761) term the 'psychological manipulation and … paranoia created by the feeling that you are always in danger'. The feeling of being 'watched' restricted her from seeking support due to those in her life 'judging her' from afar.

Uldbjerg (2021b: 4) notes that experiences of digital sexual abuse 'change over time and exposure involves not only separate instances but a recurring pattern of living with the consequences of digital assault'. The difficulty of 'living with' and indeed, after the moment of doxxing was a common burden reported by the women. The aftermath of doxxing is also another way to evaluate the gendered dynamics of shame, reputational harm, security, and well-being. Jana experienced ongoing and escalating harassment from a male colleague, culminating in a doxxing attack which included releasing her home address, workplace, and other identifying information online. Jana reported being terrified for her well-being, despite her workplace taking action and firing her colleague:

> I don't know how he got the information about me in the first place. He called me a bitch and other things I can't say now. I felt like my world was going to end. He threatened me and said that he would come to my house. The physical threat of seeing him again [at work] was resolved, but the mental impact of knowing that he knows where I live and was seeking to cause me harm lingers. (Jana)

Jana is a rare example of a doxxing victim who brought her perpetrator to some kind of justice, in this case, through the termination of his employment at their workplace. However, Jana's ongoing distress at his knowledge of her home address and indeed, her concern at how he came to obtain this information in the first place, illustrates the difficulty of finding true forms of redress for doxxing. Once personal information is disclosed, it is extremely difficult for the victim–survivor to regain the integrity of anonymity that they once enjoyed. While Jana did not engage with the police, her frustration at not being able to truly recover control over her information reflects broader challenges of legislating against doxxing through the criminal justice system.

Absent and conditional family support

Family relationships can inhibit or provide for relations of care in the aftermath of harm. Of particular focus in this section is the implication of children being ancillary victims to the doxxing of their parents. Privacy abuse and its impact on children has sparked a growing literature on the need to protect the privacy rights of children online. As children grow up with increased technology and media integration in their daily lives, academics and policy makers fear that this exposure does not extend to digital literacy in relation to data security. The popularization of family accounts on YouTube, TikTok and Instagram, depicting the everyday lives of young people and their parents, has called into question the rights which children might be denied when a parent is 'oversharenting' their sensitive and personal information to a large and unknown digital audience (Klucarova and Hasford, 2023). As Steinberg (2017: 839) notes, 'parents act as both gatekeepers of their children's personal information and as narrators of their children's personal stories', which can position the personal data autonomy of children and young people to be vulnerable to doxxing even before they develop their own desires and interest in self-presentation online.

Exploring how children are impacted by doxxing is a complex, yet important, task. Children who experience 'vicarious data harms' stemming from their parents' doxxing are often overlooked as secondary victim–survivors, where the harms to their data autonomy become an extension of their parent's harm experience (Ugwudike et al, 2024: 1). Observing how family structures can support victim–survivors – or, in some cases, construct the settings where harm occurs – requires a reflection on the experiences of children as individual and separate from the adults in their lives.

Molly's experience of being doxxed was intertwined with the doxxing of her young daughter. Molly experienced targeted, malicious doxxing from an unknown perpetrator, who used a fake account to leak her home and work addresses, her work ID, and other confidential workplace information as well as the identity and school address of her child. The immediate consequences of being doxxed in this manner were severe. Molly faced repercussions at work, in addition to being harassed by unknown callers. However, Molly's primary concern was the safety and well-being of her child, putting her needs before her own desire to seek justice:

> If my child wasn't involved, I don't know how I would've operated with the case, if I would've gone ahead with [making a formal report to the police] or not. I don't know what, I sure know that I was depressed. Most times the shame and the guilt would not make me react. Maybe I would've reacted properly if my child was not involved. But if you are a mother, you would understand that if a child of yours is involved in any case with you, there are ways you'd wanna operate with it to avoid it being a stigma on her or affecting her in the future. (Molly)

Molly considered reporting her doxxing to the police but did not proceed out of concern that the investigation could further traumatize her child, who was already depressed and 'afraid to go to school'. Elsewhere in the interview, Molly expressed profound guilt and shame that her daughter had been implicated in her doxxing and felt that she had 'let her down'. It is important to note that Molly is, of course, not to blame for her daughter's doxxing – she herself was also victimized by a targeted and malicious campaign to undermine her professionalism and intimidate her in her own home. Regardless, Molly demonstrated an effort to take back control where she could by not going forward with a police complaint. By deciding not to pursue a formal police investigation, Molly made efforts to reclaim her identity as a mother who could protect her child by any means necessary.

While there are several initially obvious instances of violence in Molly's account, such as silencing, intimidation, and harassment, other implicit forms of violence are also evident. Taking a feminist and intersectional approach illustrates how 'those marginalized by multiple structures of oppression and who are frequently most in need of support face considerable challenges in having their voices and stories heard' (Wheildon et al, 2022: 1701). Molly reflects that her family's religious and cultural judgement of her pregnancy resulted in pre-existing alienation from the family. Molly is also a single parent and has sole custody of her child, which resulted in limitations in her support network:

> I reached out to two friends, not to my family. I had my child out of wedlock. They used to use that against me and use it as an excuse not to come through for me when I'm in trouble. So I just decided that I do not have any family. So I just called a few friends to help, who helped me get my child back to school. (Molly)

Before her doxxing, Molly was already experiencing precarity as a single parent alienated from her family. Being doxxed, therefore, did not tip Molly *into* alienation from her family but rather exacerbated pre-existing 'structures of oppression' through the *absence* of familial support that she might have otherwise drawn upon. Rather than visualizing the violence of doxxing in this moment as a swift rupture, we might consider that this longstanding familial alienation is a form of slow violence. Expanding on Nixon's (2011) initial framing of slow violence, Barnwell (2019) suggests that families facilitate slow violence through secret-keeping and stigmatization. This has two effects: 'families experience a double form of violence, where the risk of social stigma pressures the family unit into silencing, or even ostracizing, family members, and then, in a second blow, leaves subsequent generations to deal with the often hurtful or confusing legacy' (Barnwell, 2019: 1112). For Molly, the ostracization from family support underscored the experience of precarity brought on by being doxxed and flowed on to her daughter too. While Molly relied on other relations of support offered by friends, her alienation from family placed parameters around the feeling of precarity and 'aloneness'. The 'unspectacular event' of a family disagreeing over values – in this case, having a child prior to being married – had a rupturing impact on Molly's social stability, 'wielding social power from within families and across generations' (Barnwell, 2019: 1124). This social power structured the roles which Molly played – a single parent caring for her daughter alone – and relegated her family outside of the relations of care that Molly needed in the aftermath of being doxxed.

In contrast to Molly's experience of familial alienation, other participants reported that family identities and relationships were integral to receiving care and support after being doxxed. Jessie and Aisha reported that their respective identities as mothers and daughters shaped possibilities for support. However, these relations of care were not unconditional but contingent on cooperation with norms and values of 'appropriate' social relationships. As the following interview excerpts illustrate, these contingent relations of care are upheld and maintained by the keeping of secrets. Secret-keeping, as Barnwell (2019: 1113) notes, is 'not always the result of individual intention – they register an intimate action, but also the pressure of a wider context that directs codes of conduct and silence'. Jessie, who experienced doxxing by having her social media account hacked and intimate photos posted, framed secrecy as a strategy to reclaim control over her information:

> I understand that trying to explain [doxxed intimate photos] to people could cause me more harm. It's something that happened and now it is something that has to stay for a lifetime. Only my children, my family get to know the truth about it. (Jessie)

For Jessie, one of the most serious impacts of being doxxed was her account being hacked. Not only did she have to grapple with the non-consensual disclosure of her intimate photos, but she also felt helpless in convincing people that she had not posted them. Jessie reported that it was 'quite difficult to convince people' that she was not behind the posts, an experience which further distanced her from support that might have been more readily available if she had been less ambiguously doxxed. Refusing to explain how her intimate photos were doxxed afforded Jessie some small measure of peace in the aftermath of being doxxed. Her silence here is not just 'the absence of voice'; but an 'interactive process that responds to the conduct of other human beings' (Roberts, 2000: 343). Jessie made the judgement that the truth was something that could only be entrusted to a close-knit community – her children and immediate family. Ahmed (2010: xvi–xvii) reflects that to 'keep a secret is to preserve a bond with the one who has entrusted us with it'. For Jessie, secret-keeping and silence became a mode of resistance against the rupturing of social bonds, where she instead reinforced relations of care and trust with her children and inner circle of family. These strategies of resistance through withdrawal and 'opting out' are discussed in further detail in Chapter 6.

Aisha's experience of being doxxed by her ex-partner was previously discussed as an example of the intersecting harms of intimate partner abuse, image-based sexual abuse, and victim blaming. Aisha's experience also highlights that female honour is mired in the slow violence of patriarchal values and family relationships. This slow violence, much like Molly's familial alienation, pre-dated the instance of doxxing. The slow violence of patriarchy underscored limitations of care, support, and trust that Aisha could receive. Like Jessie, silence and secret-keeping were strategies that afforded Aisha some comfort and control over her situation:

> I was lucky enough that my whole family didn't get to find out about it. It was just my mom. I told her because I really needed support, mental support. So she was the only one that knew about it. (Aisha)

Aisha explicitly noted that she was glad her father was not aware of her doxxing. Additionally, she stressed that she did not blame her ex-partner for doxxing her intimate photos. Rather, she blamed herself and conveyed a deep sense of shame for her sexuality and intimacy with her ex-partner. As a young religious woman in her early twenties, Aisha was keenly aware of the power that her family, particularly her father, held over her in upkeeping their family image. While many factors can determine the role of parents in supporting their children into adult life (see Swartz et al, 2011; Sage and Johnson, 2012), the variety of ways that family systems either bond together or rift apart after harm are connected to the family's value framework. As such, the relational bonds between parents and children, particularly the youngest child, in conservative and religious families are often rooted firmly in hierarchies of paternal 'protection' (Heath, 2019). For Aisha, her family framed her youth as vulnerability, particularly by denying her capacity for sexual agency and determination. This denial echoes what Albury (2017: 714) terms the view of young, typically heterosexual, women who have experienced image-based sexual abuse as 'lacking agency or sexual rights while bearing individual responsibility for the public exposure (or "consequences") of sexual self-representation'. Additionally, the cultural context of religious conservatism also situated Aisha as a 'risk' to the family's honour, calling in the need for her father and male relatives to protect their family's reputation and social standing. As Christianson et al (2021: 1) note, societies with honour codes treat the protection of the family as a responsibility of male family members, extending 'beyond physical and economic protection as control of a female's sexual behaviour'.

While Aisha's complex relationship with her family is shaped by culture, religion, and conservatism, it is important not to silo victim blaming as a uniquely cultural or religious problem. Patriarchy is evident across nearly all cultures and organized religions, although postcolonial feminist scholars have noted that western modalities of feminism 'outsource' the problem of patriarchy to the Global South by calling for the 'saving of brown women from brown men' (Spivak, 1994: 92; Grewal, 2013). Moreover, patriarchal societies 'condition people to focus on the exoneration of male perpetrators … by pivoting women as the cause of and the solution to all male violence, instead of focusing on the male committing the offence(s) against women' (Taylor, 2020: 56). Across the interviews, many women of varied religious, cultural, and ethnic backgrounds reported similar condemnation and shaming for having their intimate images and videos doxxed. Across these diverse experiences of doxxing, however, is how honour is ruptured when non-consensual disclosure causes the victim–survivor to 'fail' to uphold their status within the hierarchy of worth.

Justice, law, and police responses in the aftermath of doxxing

Legal responses to doxxing are limited, inconsistent, and largely unclear to victim–survivors in many jurisdictions. McGlynn, Rackley, and Houghton (2017: 31) highlight that legal responses risk 'shoe-horning practices into conventional privacy-related offences', which in turn 'risks obscuring the nature of the abuse'. Globally, law enforcement agencies continue to come under scrutiny for poor responses to technology-facilitated violence and abuse (see Flynn, Powell, and Hindes, 2023). While there have been some amendments to laws which criminalize types of digital harms and data abuse in Australia and the United Kingdom, where legislation which criminalizes doxxing exists, it is typically not common knowledge. When asked whether they had reported their doxxing to law enforcement, only three participants had made a formal complaint. Even among those participants who had made a complaint, police declined to investigate their complaints. These barriers highlight the challenges of not only knowing *how* to seek justice in the aftermath of doxxing by making a complaint but a deeper issue of *recognizing* that doxxing harms might be captured in obscure or arcane legal frameworks.

Among the women, knowledge of legislation criminalizing data abuse and reporting protocols was low. This is hardly surprising given the range of legal mechanisms which do – and do not – address doxxing as a specific criminal offence. While some jurisdictions – such as the Netherlands, Australia, Indonesia, Hong Kong, New Zealand, and South Africa – have explicit laws criminalizing doxxing, many other countries indirectly respond to data and privacy abuse. The unclear legal status of doxxing as a criminal offence impacts the pathways that are available to victim–survivors as they negotiate the ongoing threats to their privacy and data security. When asked about the criminal implications of doxxing as a form of data and privacy abuse, most participants were not aware if doxxing was criminalized in their jurisdiction. Hara expressed incredulity that the police would take her complaint of doxxing seriously:

> I wasn't even aware of any law against doxxing ... Since it was something that was trending on social media, I didn't think like the police would help or they were going to really do something about it. (Hara)

Hara's concerns reflect ongoing challenges with police resources, training, and internal police cultures which lack insight into victim–survivor needs after experiencing online harm. Australian police responses to digital abuse have been critiqued for poor response times, lack of understanding about doxxing legislation and inadequate resources and training of personnel. Wilson et al (2022: 487) found that only 38.35 per cent of general duty police officers had completed cybercrime reporting protocol training, with 18.05 per cent also being trained on the preservation of digital evidence. Although Australian cyber policing divisions are in the process of upskilling officers and assigning more resources to the training and administration of investigating digital abuse (Cross et al, 2021), the trickle-down benefits of these changes are not yet evident in victim experiences of reporting doxxing to police.

Victim–survivors of all kinds of gender-based violence often report that engaging with police can retraumatize them in the aftermath of harm. The lack of compassion and insight into the trauma endured by victim–survivors has been identified as a key area of concern for major policing bodies, including the National Police Chiefs' Council (2024: 38), who highlighted that trauma-informed processes were a priority in their violence against women and girls reform agenda. Where policy frameworks are in place to criminalize digital harms and abuse, they often fall short of considering the interpersonal experiences of victim–survivors. In her review of the Republic of Ireland's *Harassment, Harmful Communications and Related Offences Act* (HHCRO), Hayman (2024: 11) critiqued that it 'lacks a victim–survivor-centric approach to harm, fails to reflect a (digitally) embodied understanding of harm, and does not approach harm from an intersectional perspective'. These limitations in legislative and policy responses to serious breaches of sensitive, personal information create a justice landscape which is inaccessible, out of touch, and unresponsive to the needs of victim–survivors of image-based doxxing.

When police personnel are not equipped to engage with victim–survivors in a trauma-informed manner, victim–survivors may be deterred from reporting or seeking support in the aftermath of harm. This is evident in Odette's case, where her fear of being shamed deterred her from reporting her doxxing to local law enforcement:

> Why would I think that the police would be by my side? They would also shame a grownup woman like me before taking action. I didn't want any more emotion, I didn't want any more guilt. I felt accused of everything. I felt accosted. (Odette)

Feeling 'accused' in the aftermath of experiencing privacy abuse is yet another example of how informational autonomy is ruptured by doxxing. Importantly, this kind of rupture requires a parallel negotiation of composite parts of the self, where the harms to data security and autonomy become crowded out by the intensity of the shame and blame arising from the disclosure. This kind of splintering – elsewhere recognized as compartmentalization in complex post-traumatic stress scholarship (Clifford, Hitchcock, and Dalgleish, 2020) – reflects what Brison (2017: 224) described as the challenge of 'remaking the self' in the aftermath of trauma. When victim–survivors of doxxing choose to report their experience to police, they are at risk of not being believed, feeling shamed, and being framed as the 'guilty' party. This signposts that police knowledge and training of trauma-informed processes are a key barrier to overcome to ensure that victim–survivors can make criminal complaints in a manner where they are supported, validated, and believed.

Participants also faced significant challenges in interpreting arcane legislation relating to doxxing, regardless of their location. Four of the women volunteered that they were located in Australia, where doxxing is criminalized under the *Criminal Code Act (1995), (s 474.17)* as the 'use of a carriage service to menace, harass or cause offence'. s 474.14 does not specify any particular form of cyber enabled abuse, but is used as an umbrella clause for any type of online harassment with intent to menace, harass, or offend. As Powell and Henry (2017: 210) have noted, this 'law has rarely been used in relation to these types of behaviours'. At the time of writing, no criminal charges against doxxing offences have been successfully prosecuted in Australia under s 474.14 of the *Criminal Code Act* (1995), although new 'doxxing laws' were passed in the *Privacy and Other Legislation Amendment* in late 2024. 'Carriage service' framings of legal liability are insufficient in capturing what Jose (2017: 2) terms the 'scalability' of the internet, where acts like doxxing can 'effect maximum harm with minimum effort' from perpetrators. The hesitance of participants in this study, as well as their lack of confidence in accessing and interpreting legislation criminalizing doxxing, plays out as an inhibition of their ability to seek justice.

Doxxing harms illustrate the need to move beyond the limited imagination of a singular perpetrator accessing a carriage service to harass, offend, or menace in a singular moment of harm. Most participants described their doxxing experience as going viral, with multiple reshares or engagements with the post, each of which could constitute a separate form of harassment in some jurisdictions. In Tamara's case, where her intimate images were sold to multiple anonymous parties, the initial disclosure of her personal and sensitive information would comprise a first offence, and subsequent reshares of the doxxed material comprising additional offences. This is increasingly complex when doxxed information might fall under multiple legal categories – for example, image-based abuse and exploitation material. While it is promising to see some jurisdictions across the globe revisit and amend their existing privacy and cybercrime legislation, further nuance is needed to capture the complex, imbricated realities of personal information – and the 'essentially undeletable' vulnerabilities that doxxing exposes victim–survivors to across multiple modalities of their identities and social environments (Romania, 2022: 357). Global legal recognition of doxxing as a criminal offence is an important step towards addressing the cyclical nature of informational abuse. This also echoes Corbridge's call for better legal options and remedies in response to doxxing, moving toward the right to informational self-determination through the regulation of personal data protections (see Corbridge, 2018; EU General Data Protection Regulation). While legal reforms are an important next step, they must also be affirmative as well as prohibitive. Legal change without recognition for the informational autonomy of all people will form an incomplete picture of the enduring, relational harms of doxxing.

Conclusion

The women's experiences discussed in this chapter have revealed the complex intersections of gender, shame, worth, and honour: factors of identity which are uniquely challenged in the aftermath doxxing. These accounts offer important insights into the specific ways that gender influences both the *form* of doxxing – such as retributive, targeted, and malicious forms of doxxing through digital sexual abuse – and the *type* of harms which arise from it. The women interviewed in this study reported long-term and serious impacts to their mental health and well-being, their success at work, and their connection to family, friends, and their social worlds. These harms often took many years to heal, if indeed they could be recovered from. Situating the harm of doxxing endured by women as a form of sexual shaming and disavowal is a key finding from this chapter. The shame invoked through doxxing positioned the women within the patriarchally defined 'hierarchy of worth', a hierarchy which offers little sympathy, care, or support for those who are situated at the bottom.

Many women who are doxxed are targeted through image-based sexual abuse, digital sexual assault, and the exploitation of other vulnerable and gendered information, such as their reproductive and sexual histories. It is no coincidence that these forms of information relate to women's bodies, sexualities, and identities. This illustrates the ways in which patriarchal value judgements structure and influence gendered forms of violence, including gendered privacy and data abuse. This chapter has shown that doxxing reproduces and compounds harms in digital and mediated spaces. While image-based abuse scholars have noted that perpetration of this form of digital assault may comprise doxxing, I underscore that doxxing functions to *mediate* image-based sexual abuse, while simultaneously *compounding* it through the additional risks to privacy, security, and personal information. This framing moves away from considering doxxing to be a technique to produce 'other' forms of digital abuse and technology-facilitated violence. Understanding image-based sexual abuse, sextortion, and other forms of digital gender-based harassment to be intersecting with doxxing, rather than merely facilitated by doxxing, reflects a growing pivot in technology violence and abuse research to consider how technologies 'invite, afford, and translate' harms, instead of just facilitating it (see Wood et al, 2023; Ugwudike, 2024).

In this chapter, I have taken up the language and stylings of gender in my participants' own words. At times, this accounting put forward a restrictive and binarized account of gender, relationships, family dynamics, and other insights into gendered social life. When these moments crop up, I contextualized this as an exemplar of patriarchy in the ways my participants understand themselves and their social worlds. This is not to suggest that women experience a singular 'knowable' kind of doxxing, and men another kind. An important consideration is that this chapter reflects only on the experiences of cisgender and heterosexual women and men, as interview recruitment did not attract participants who are gender diverse, queer, or otherwise existing outside of a binary conceptualization of gender. While the importance of researching with gender-diverse people cannot be understated, it is telling that cisgender and heterosexual participants had the vocabulary to self-ascribe their experiences as doxxing. This reflects a recognition of doxxing as a technique of harm and abuse within heterosexual relationships, which often model and comply with patriarchal value judgements.

Accounting for the violence of doxxing as a rupturing of social life may seem bleak. Ruptures through doxxing might indicate that informational autonomy, once lost, cannot be regained. Rather than framing doxxing in these pessimistic terms, accounting for the violence of non-consensual disclosure as the rupturing of relational ontologies might instead present the tools and possibilities for reclaiming autonomy over personal information. Just as doxxing harms coalesce in the social, so might social bonds of care present opportunities to reclaim, resist, and amend the harms of doxxing.

5

Overcoming Doxxing: Masculinity and Ruptured Honour

Introduction

Privacy abuse is neither static nor linear, and it is experienced socially in myriad ways. In this chapter, I outline how men, who experience doxxing within more defined social networks of peers, have access to different support mechanisms and opportunities to 'overcome' their doxxing. This highlights how non-consensual disclosure plays out in gendered ways, reifying the hierarchization of worth of men over women, children, and gender-diverse people. The men reported being doxxed through embarrassing video footage and social media 'call-outs' which accused them of abusive and violent behaviour towards their partners. Unlike the accounts of the women, most of the men reported being doxxed within their immediate social groups, rather than their doxxing going viral to a larger number of people. As such, their humiliation of being doxxed played out within homosocial contexts, creating repercussions to their social standing among peers, friends, colleagues, and family members.

The men's experiences of being doxxed were also taken seriously by their families, and they were subsequently provided with different offerings of support and care. While support was more freely offered to the men I interviewed, their accounts of doxxing still invoked shame, social rupture, and concern for their future in similar ways to the experiences of women. By 'falling short' of the hegemonic ideals of masculinity, male victim–survivors of doxxing also became positioned lower in the hierarchy of worth (see Westmarland and Burrell, 2023). I discuss that in the aftermath of doxxing, the men felt that they had to demonstrate 'strength' to prove the 'type' of man that they were – one who could overcome and move beyond the rupturing provoked by doxxing (Idriss, 2020). Bringing honour into conversation with strength shows the limited ways that men could express and experience victimhood after doxxing. This illustrates how patriarchal systems of relations encroach on possibilities of affirming, expansive care, and respect.

In the previous chapters, the shame invoked through doxxing women's sexual histories and bodily information highlighted how doxxing disclosures uphold patriarchal values of worth. Shame shows the pressure points of our social lives, illustrating where acceptance and support begin and end. As Barrett (2018: 38) notes, 'it is not only that the shamed individual is imagining herself or himself from the perspective of another, but rather that (s)he is operationalizing a shared universe of moral judgments, status hierarchies, and social expectations to evaluate the self'. This chapter will flesh out some of the 'status hierarchies' that inhere within the gendered differences of doxxing as reported by participants in this book. This calls into focus a key distinction of the violence of social rupture: victims of doxxing are differentially judged through the imposition of moral hierarchies in the aftermath of non-consensual disclosure. In this way, violence is imposed through the norms and ideals of gendered hierarchies of worth. As such, shame is integral to understanding the violence of social rupture, as it illustrates the ways that these norms are discriminatorily and differentially applied to victim–survivors along gendered lines. Shame also has highly gendered ties to masculine conceptions of honour, status, and social standing.

Appeals to honour, whether it be to honour lost through doxxing, or honour regained through support or 'strength,' underscored the experiences of almost all the men interviewed in the study. Honour within patriarchal social systems has been taken up as a moral foundation, to describe the values of chastity, integrity, and strength (Idriss, 2020). As such, honour has deeply gendered connotations about who can achieve it and who must comply with it. This chapter takes inspiration from Raewyn Connell's framing of honour as a dividend of patriarchy, achieved when men are in compliance with hegemonic ideals about masculine strength, dominance, power, and control (Connell, 2005a, 2005b, 2018). As Ozyegin (2018: 245) notes, dividends are not always pursued intentionally by men but are produced through 'the complicity of other, subordinated masculinities (despite their deep contradictions), who benefit from the patriarchal dividend'.

Honour has a particular connotation in criminological and sociological research into men's violence toward women and other men. 'Honour-killing' or 'honour-based violence' literatures have framed honour as something that patriarchal, religious, and cultural systems use to justify men's violence against women (Rowlands, 2022). These framings highlight the enmeshment of honour with male 'norms which equate strength in men with dominance over women' (Katz, 1995: 163), but also note that honour underscores male violence toward their peers (Tomsen and Gadd, 2019) and 'provocation' to homophobic violence (Tomsen, 2017: 824). A recent shift in scholarship has recognized a wider range of honour-based violence and its victims (Idriss,

2020), including research into the experiences of queer people and people with learning disabilities being coerced into forced marriages (Khan and Low, 2020; Dutt, 2020; Clawson, 2020).

Honour has long been brought into conversation with research into violence. While this chapter takes these framings into account, I focus on how honour is operationalized to different effects and outcomes by the men and women in this study. Honour, I argue, is key to understanding the nuances of gendered experiences of doxxing, and, more broadly, shows how the social rupturing of honour affords men and women different possibilities for support, reclamation, and recovery in the aftermath of doxxing. I am particularly interested in bringing these differences into conversation with Braidotti's (2019c: 36) explanation of the consequences of 'ceasing and becoming', where doxxing harms play out in both the present and future timelines of victim–survivors.

In this study, men experienced doxxing as a transgression to their honour which breached codes of peer-to-peer friendship. Women, contrastingly, experienced doxxing as disclosures which amplified 'dishonourable' information about their bodies, medical histories, and sexual identities. As such, the violence of non-consensual disclosure is structured along the gendered lines of patriarchy. Women's experience of doxxing harms are funnelled through the lens of patriarchy, where disclosures can dishonour and sully not only their relationships but their broader social standing in their families and communities. While men experience the harm of doxxing as a form of humiliation, having prior access to the dividend of honour gives them different strategies to reclaim or restore this status. Therefore, the timeline of experiencing harm is framed on a different axis – men experience doxxing as a harm that they can 'overcome' through 'strength' and with family support, whereas women experience doxxing as a harm that sticks with them 'forever'. Dishonour is located either within the self or externalized to social status and standing.

This chapter does not suggest that male experiences of doxxing are less serious or significant than female experiences of doxxing. Rather, it points to the ways that doxxing is experienced in different social contexts which are shaped by gendered influences and socializations of different victim–survivors. In the previous chapter, I discussed how patriarchy and misogyny give shape to the forms of familial alienation and sexual shaming experienced by women victim–survivors. In this chapter, I again consider how patriarchy structures male experiences of doxxing not only in the social form that it takes but through fostering expectations that male victim–survivors will 'overcome' and 'grow stronger' in its aftermath. I also note the paradox that harm and strategies of support coincide within 'homosocial' relationalities, which both sustain the humiliation of doxxing but also offer strategies of support and recovery through male friendship, intimacy, and care. This paradox highlights what Haywood et al (2018: 73) determine as the ways that 'different masculinities correspond to (and underpin) different ways of articulating hegemonic masculinity', and broaden out understandings of how patriarchy harms beyond its maintenance of hierarchies between men and women.

Social humiliation as an affront to honour

The men interviewed in this study experienced different types, outcomes, and implications of doxxing compared to those reported by female victim–survivors. In this context, doxxing was a harm that registered socially, either in the form of humiliation, competitiveness among peers, or through repercussions in their romantic, platonic, and workplace relationships. Doxxing was also taken seriously by the men's families, and they were subsequently provided with different opportunities for support and care. While support was more freely offered to the men I interviewed, their accounts of doxxing still invoked shame, social rupture, and concern for their future in similar ways to the experiences of women.

Four of the five men[1] interviewed in this study described their experience of doxxing as one of social humiliation. Notably, the doxxed information did not pertain to sexual images or videos, as was overwhelmingly the case for the female interview participants. Doxxed information included the release of sensitive information about their intimate relationships, allegations of perpetrating abuse within their intimate relationships, as well as video recordings of intoxication, violence, or impersonations they found to be humiliating. Unlike the accounts of female participants, most of the men described their doxxed information playing out within smaller, more immediate social groups, rather than going viral to a larger number of people. As such, the resulting humiliation of their doxxed information played out within a close social proximity, creating repercussions in their social standing among peers, friends, colleagues, and family members.

I distinguish the men's experience of humiliation from the women's repercussions of shame after doxxing using Katz's (1988, 1989: 358) argument that 'shame requires the internal acceptance of affront', whereas 'humiliation must be transformed into rage'. As Chapter 4 demonstrated, women experienced sexual shaming through doxxing as a consequence of their diminished position within the hierarchy of worth. Many of their experiences evidenced a degree of self-shaming, where they regretted or blamed their previous sharing of personal and sensitive information with ex-partners who later exploited it. In this chapter, I will explore how men conceptualize the humiliation of their doxxing as an affront to their masculinity which they refused to accept and respond to through demonstration of strength and overcoming, a process which attempts to realign them with heteronormative performances of masculinity.

Joe, a young man in his late teens, experienced social humiliation via the doxxing of a video in which he imitated a cow eating grass. As Katz (1989: 359) notes, 'in humiliation, one is totally opened up to social forces which make one powerless'. At the time of this event, Joe was still in high school and experienced prolonged harassment and bullying from his classmates. Joe was accidentally doxxed by his younger brother, who had filmed him trying to care for a sick cow on a family member's farm by imitating eating grass in an attempt to encourage the cow to eat. According to Joe, the video had been innocently captured but 'fell into the wrong hands' of his brother's friends, who shared the footage on social media:

> I started seeing some of those [videos] on social media with my friends in school. And it was pathetic. It was terrible. I was very, very, very shocked. How can this happen? (Joe)

Joe's shock at his classmate's reaction demonstrates an interesting aspect of how doxxing harms are mediated and specific to the context in which information is shared. Had Joe's classmates seen the video and interpreted it as an act of 'caring for a sick animal', as Joe intended, it seems less likely that he would have been humiliated. He might have felt pride, or felt 'heroic' for coming to the aid of someone in need (Frisk, 2019). After being humiliated by his classmates, Joe described the video as a form of private, sensitive information, which he might not have done if he had been praised for his efforts to help the cow. This reflects what Nissenbaum (2004: 137) terms the 'texture of people's lives', where the context of our information, and our subsequent framing of that information as private or sensitive, not only '[crosses] dichotomies, but [moves] about, into, and out of a plurality of distinct realms'. Joe's feelings about his impersonation of the cow depended on the varied ways in which it could be interpreted socially – and his reaction to the video being doxxed thusly 'moved about' into the realm of humiliation.

This is a significant departure from the types of doxxing that women reported in this study. In contrast to the sharing of personal and sensitive information about sexuality, bodies, medical histories, and other gendered information, Joe's experience appears to have been interpreted as harm due to its reception by his peers. While I have briefly noted unintentional doxxing in earlier publications (see Anderson and Wood, 2021), there is a gap in the literature engaging with the types of information and harm that might emerge from this form of doxxing. Nonetheless, Joe reflected at length that the experience caused him anxiety and distress, and impacted his social well-being. Here, the fallout of doxxing might be conceptualized by what Ozyegin (2018: 244) terms 'the desire for recognition', where the 'uncertainty of rank and porous dividing lines between different types of masculinities' shaped Joe's social humiliation and denigration by his peers. This is one example of how patriarchy harms men and boys, in addition to women and girls.

By 'falling short' of the hegemonic ideals of masculinity – namely, by showing 'weakness' or 'silliness' by imitating a cow – Joe was positioned as lower in the hierarchy of worth by his classmates and friends. As Marchia and Sommer (2017: 271) note, 'gender performances other than hegemonic masculinity and idealized femininity are considered abnormal under the patriarchy, which constructs these norms', which explains Joe's distress at his actions being interpreted as embarrassing rather than helpful. Acting in a manner perceived to be 'abnormal' to the performance of masculinity demonstrates how heteronormative values shape the judgements that arise in moments of doxxing. It is also worth noting that deploying humiliation as a technique of domination shows only one facet of patriarchal systems of relation, rather than a universal form of patriarchy. Interactions between adolescent boys cannot – and should not – be extrapolated to represent relations of patriarchy across all contexts. Ozyegin (2018: 250) calls for a theorization of patriarchy which '[incorporates] a focus on interaction/intersection with other systems of domination', which here might include the intersection of class (as shown by working on a farm) and age (such as the dynamics of adolescent socialization at school), in addition to the patriarchal 'one-up manship' which can motivate targeted humiliation. As such, Joe's experience addresses a specific and contextualized form of social rupture where patriarchal relations structured many axes of his identity as a young man.

Payback, control, and challenges to masculine identity

Most of the men described their doxxing as being motivated by payback or retribution. Retributive doxxing is 'underpinned by a perpetrator's motivation to punish a victim through facilitating harassment against them', commonly deployed to incite repercussions at work, at home, or in other social contexts (Anderson and Wood, 2021: 214). David, in the following excerpt, described how payback-doxxing resulted in him losing control over his social status:

> They are doing it to get at you; doing it more like a payback – this is something that you don't have control over, irrespective of how careful you try to be in life. (David)

David's description of trying to 'control' and be 'careful' about his reputation reflects what Tomsen and Gadd (2019: 18) term 'retaining 'face' as a means of cultivating 'respect and social honour'. In this theorization of honour, masculinity hinges on the maintenance of relations of respect within patriarchal hierarchies, taking the form of acknowledgement from other men of the same 'status', and the acquiescence of women, children and others perceived to be 'lower' in the hierarchy. Payback, therefore, is dually harmful because it denies the maintenance of respect while also being the conduit for harms like doxxing which lower status within this hierarchy. Across interviews with both the men and women in this study, humiliation was instrumental to the motive of payback. As Andrew reflects, humiliation as a technique of payback is highly intentional, representing a targeted motivation by the perpetrator. Andrew was doxxed by a university classmate who he perceived to be 'in competition' with him for the attention of his girlfriend. The doxxer filmed and shared a video of Andrew at a party in which he was highly intoxicated and acting erratically. In his words, Andrew perceived his doxxer to be egging him on:

> He was laughing with me. He was doing, even making me say so many things. I think later he went, he hid himself and took the video apart from him, nobody had the video. He was the only person that had the video. (Andrew)

When the video was shared, Andrew experienced repercussions among his university peers, as well as in his relationship with his girlfriend, who broke up with him after watching the video. Andrew was mostly concerned with the social repercussions of his university friends who had seen him behave erratically and violently in the video. At one point in the interview, Andrew described that attendees at the party had to intervene to de-escalate his physical violence. While he noted that he regretted his escalation into violence, Andrew shifted blame onto the perpetrator for 'making me say so many things' that he later regretted, and accused him of editing the video to 'make me look bad'. This shifting of blame maps onto what Tomsen and Gadd (2019: 26) describe as the matching of 'violence with shifting, but seriously felt, views about respect, justice and fairness in the treatment of other people'. Andrew framed his escalation into violence as a regrettable but understandable response to his perception of unfair treatment.

Similar to Andrew's account, David experienced the repercussions of doxxing within a social context where his status mattered. One of the only male victims to be doxxed by an anonymous perpetrator, David was doxxed when allegations about him abusing his girlfriend spread on social media. David was extremely hesitant to describe the nature of the allegations in the interview, expressing concern that these details might identify him. It is not the focus of this analysis to appraise the veracity of David's claims of innocence but to consider how informational disclosures might be utilized by perpetrators to conceal their abuse and use of violence. Deny, attack, and reverse victim and offender (DARVO) literatures have found that many perpetrators reframe their actions as 'appropriate' to misdirect accountability for their abusive behaviour onto the true victim (Harsey and Freyd, 2020; Kadri, 2020). David's account does, however, provide insight into how perceptions of 'weakness' figure into male experiences of social rupture. Instead of detailing the type of information leveraged against him, David focused heavily on speculating who might have shared the information. As the following excerpt suggests, David believes that his partner 'narrated the story' to a friend who took her complaint against him seriously and disclosed his alleged abuse online:

> The whole story was written in a way that made it look like I was the weaker one. I was the one at fault. And then I was the one who was treating my partner badly. That's why I had nursed the thought that it could be her friend, which she must have narrated the story too. And then they themself would then conclude that, oh, that's the real story and decide to take it up. (David)

When asked about the enduring impact of doxxing on his life, David expressed frustration that he could not clear his name. He described asking his partner to make a post on social media denying the allegations, although it is not clear whether she agreed to do this. It should also be noted that attempting to control or monitor a partner's posts on social media falls within coercive control behaviours, a form of digital intimate partner abuse (Harris and Woodlock, 2019; Woodlock et al, 2020; Cuomo and Dolci, 2021; Walklate et al, 2022). While David did not reflect further on how his partner, who he was still in a relationship with at the time of interviewing, responded to his request, the request itself shows a desire to 'right the wrongs' that his doxxing stirred up. Here, it is useful to frame the harm of doxxing as denying 'aggrieved entitlement', which Kimmel (2013: 454) terms as 'compensation for humiliation'. Thorburn, Powell, and Chambers (2023: 242) note that men and women have different avenues for pursuing aggrieved entitlement, although 'men appear more likely to discharge their aggression outwardly and violently'. In David's case, the fear of being perceived as violent chilled his ability to deny the allegations. He did not want to validate or affirm the rumours that he was an unstable or violent man. As a result, David could not perform the 'aggrieved entitlement' required to repair the rupturing that he experienced post-doxxing. Misrepresenting abusive behaviour in a research interview could be a strategy of taking back control, seeking validation, and trying to shift blame onto the victim. These strategies might also be ways that someone like David could perform 'aggrieved entitlement' to recover from the public 'humiliation' of being doxxed (Kimmel, 2013).

Interviewing perpetrators of abuse is a complex and difficult process. Kelly and Westmarland (2016: 116) have shed light on some of these challenges by drawing on Hearn's (1998) concept of 'incidentalism' – where men's violence towards women is framed as a 'one off', not that 'serious', and not really 'violence' because it is conceptualized as an individual incident rather than a pattern of manipulation, control, or abuse. Viewing violence and abuse against women as an incident rather than a pattern provides abusers with a language to 'excuse, minimize, disavow and detach themselves from their harmful behaviour' (Kelly and Westmarland, 2016: 115). Cavanagh and Lewis (1996: 87) also identify that abusers are adept at manipulating, charming and '[enlisting] sympathy' from the researcher. They urge that 'an important part of doing any work with abusive men is resisting the invitation to collude with them' or validate their explanations and rationalizations for their abusive behaviour (Cavanagh and Lewis, 1996: 99). In the following discussion of Walter's interview, these tensions loomed large. I preface this discussion by noting that at the beginning of the interview, I endeavoured to strike a balance between engaging with Walter's account in good faith, before identifying that his motivations for participating in the study were to validate and affirm his 'version' of events.

Walter, a man in his late 20s, claimed that his close female friend doxxed him by sharing historical allegations of him perpetrating a rape against another woman. Walter believed that his friend was motivated by revenge after he started a relationship with his current girlfriend, suggesting that she was romantically interested in him:

> While I was little, I had been accused of rape case, which she knows it was not true. But then she brought up this whole story and made it look like that incident that happened a long time ago was true. And then people started saying things differently. Why? Because they would say she is my best friend, and she wouldn't lie against me, but not knowing that she's doing all this as a result of me not choosing her. (Walter)

Before engaging in analysis with this excerpt, it is important to note that false allegations of rape are incredibly rare, with an estimated 0.005 per cent of allegations made to law enforcement being false or misleading, and the vast majority of survivors never formally reporting (Belknap, 2010: 1335; Lathan et al, 2023). This moment in the interview presented a significant ethical challenge to me as a researcher, as I am hesitant to platform the accounts of abusers. This was difficult to balance with my ethical commitments to engage with all participants in good faith. Debating the validity of Walter's claim of being a victim of false rape allegations is neither fruitful nor provable in the context of this study. Walter's suggestion that the doxxer was motivated 'out of jealousy' should also be treated with incredulity, as this framing also reflects DARVO tactics to shift culpability onto the victim in an effort to deny accountability (Harsey and Freyd, 2020; Kadri, 2020). What *is* fruitful, however, is the insight this excerpt gives to the social dynamics of humiliation at play when a perpetrator of doxxing does not act anonymously, as this allows the repercussions of doxxing to play out *within* contextualized social networks.

In Walter's case, the social dynamics of legitimacy underpinned how he felt harmed by doxxing. As the doxxer spoke from a place of legitimacy – the legitimacy of being not only his best friend but a lifelong family friend – Walter experienced humiliation that had no clear-cut remedy. Walter reflected that his doxxer shared 'personal stuff we've discussed together about my life, my secret things, my weaknesses'. Bringing back Nissenbaum's (2004: 137) concept of the 'textures' of context, the disclosure hinged on the prior closeness of their friendship, which doxxing ruptured beyond repair. The legitimacy of the doxxed information harmed Walter not only in his social relationships but presented a risk for future legal repercussions. Indeed, the actions of the doxxer in this case could be classified as a form of 'digilantism', a type of informal justice seeking used to 'scrutinize, denounce and even leverage harm against those deemed to transgress legal and/or moral boundaries, with the intention of achieving some form of justice' (Trottier, 2019: 197). Doxxing has been conceptualized as a form of digilantism, particularly in research examining whistleblowing, victim advocacy, and strategies of online informal justice seeking (Kasra, 2017; O'Neill, 2018; Favarel-Garrigues, 2020). While the doxxer's motivations could be seen as an effort toward digilantism, it is unclear how effective these efforts were. When questioned about the lasting impact of doxxing, Walter suggested the following:

> People were beginning to talk about that rape case, that incident, and it was more like I didn't say anything. I didn't comment, or I didn't deny this allegations. I didn't say anything about it. I just ignored it. So it was small. I believed in one thing. My mom told me that if I strongly believe that deep down I am innocent, I shouldn't let those things bother me. I shouldn't let what anybody says get at me. Why? Because my mental health is of great importance to me. (Walter)

In Chapter 4, I discussed the multiplicity of ways that patriarchy underscores social relations and contexts of worth. Walter's account shows how self-belief, validation, and familial support can be obscured into effective tools of patriarchal domination. It is particularly telling that the women interviewed in this study were condemned for their doxxing, losing work, income, family relationships, and friendships. For women like Molly, Odette, and Aisha, they could not 'ignore it' or treat their doxxing as something 'small' or insignificant. Instead, doxxing ruptured their relations to community and shattered their self-concept in ways that prevented suppression or denial. They were treated as already guilty, worthy of condemnation, and shame. These accounts illustrate the varied and complex way that doxxing ruptures social worlds along gendered lines. The default of being believed and being supported will be further explored in the next section, where honour is experienced – or denied – according to gendered hierarchies of worth.

Patriarchal dividends: the restoration of honour through 'strength' and 'overcoming'

The previous section explored the impacts of social humiliation as a key consequence of peer-to-peer doxxing between men. While these experiences of doxxing cut across a variety of social contexts, an undercurrent of honour is evident in the meaning-making of losing control over reputation. This section will consider honour as a patriarchal dividend which doxxing ruptures. Patriarchal dividends, as termed by Connell (2005a), position men to benefit from 'honour, prestige and the right to command', in addition to material privileges like wage dominance, simply by virtue of being men (Connell, 2005b: 82). As Idriss (2021: 11919) notes, 'although all men begin by benefitting from the "patriarchal dividend," not all are able to maintain it'. Framing non-consensual disclosure as a challenge to the dividend of honour casts the responses of male interviewees in a new light. Reclaiming honour after experiencing doxxing dually serves as a way of resisting, processing, and recovering from the harm of non-consensual disclosure as well as restoring the male victim–survivor to their previously enjoyed status within the hierarchy of worth.

What does it mean for men to experience doxxing as a challenge to their masculinity? Catton, Dorahy, and Yogeeswaran (2023: 8334) identify that victim–survivors often experience invalidation of their experience, creating 'disunity between the victim's experience and an accepted social expression in which to channel that experience'. This is particularly challenging for men who are victimized, as they may avoid discussing, disclosing, or identifying their experience as one of victimization in a bid to protect their self-esteem (Finkenauer and Rime, 1998: 47). Idriss (2021: 1190) also notes that validations of harm are often at odds with rhetorics of strength, suggesting that 'men who are contemplating reporting abuse may fear confirming that they are powerless, undermining dominant and social ideals on masculinity'. By reaffirming their power over their experience – by saying it 'made them stronger' – a reclaiming over their experience could take place instead. Strength cropped up thematically across many of the interviews with male participants, under a variety of guises. As David infers, strength (or its absence) is characteristic of the hegemonic ideal of 'being a man':

> I just had to mature the mindset that the person wasn't bold enough to approach me. That person, he or she's doing it to get at me, to me to pull me down. That if that person wasn't man enough to come to me directly instead of doing things, then he or she knows it wasn't right. (David)

In this accounting, David conveys that being 'man enough' is synonymous with coming forward, being direct and taking responsibility head on. Connections between masculinity and strength have been well evidenced in the literature (see Connell, 2005a, 2005b, 2018). However, the framing of strength as evidence of being 'man enough' subscribes to a patriarchal accounting of status and worth. As Seidler (1997: xi) notes, limiting the framing of masculinity as being 'man enough' '[minimizes] what men [can] learn from feminism and gay men'. This calls back to what Ozyegin (2018: 244) describes as the 'porous dividing lines between different masculinities', where a strong man is defined through action and a weak man is defined through the absence of action. Strength in the aftermath of doxxing, therefore, requires male victims to stake a claim on the 'type' of man that they are – one who can overcome and move beyond the rupturing of their honour provoked by doxxing.

Other responses indicated that complying with honour codes, religion, and other forms of respectability were also at the heart of resisting the harms of doxxing. Earlier, I noted the influence of religious beliefs and value judgements about respectability and modesty as ways that women experience shame through doxxing. Maintaining the patriarchal status quo through religious values was similarly at play in the experiences reported by the men. Andrew described his father's motivations for supporting him as a means of upholding his religious identity:

> My dad is a very respected Christian, so he's just trying to make it go away, and not try, trying to make things go very viral. (Andrew)

As discussed earlier in this chapter, Andrew was doxxed when a university peer captured and shared a video of him in a drunken tirade at an event. In the latter part of the interview, Andrew revealed that his parents helped him to move overseas as a means of coping with the consequences of his doxxing. This is a starkly different family response when compared to the conditional support – or more often, the absence of support – from family that the women participants reported in their interviews. However, Andrew's access to family support is still embedded within a strict honour culture, and is therefore contingent on his compliance with hegemonic codes of honour and respectability. While Andrew spoke favourably about his international move, it should be noted that research into honour societies reflects that tactics of control and domination underpin the movement and displacement of family members who betray the honour of the family. Men and women who are forced into arranged marriages are often required to move to different cities or countries by their families (Clawson, 2020; Mosawi, 2020). By trying to 'make it go away' – in fact, quite literally *sending* Andrew away – the status quo of respectability and honour remained undisturbed. This also calls in a comparison between the experiences of the male and female participants of this study.

While Andrew could escape and mitigate the rupturing that his doxxing provoked for him and his family, most of the female participants experienced ruptures which targeted their social standing in material and embodied ways. That is to say, the women in this study experienced the dishonour as something *located within their bodies* – be it in the form of doxxed intimate photos and videos, information about their reproductive histories, or in the first-order threat to their physical security when their home and work addresses were leaked. Whereas for men like Andrew, doxxing ruptured their honour in social contexts that, with distance and time, could be regained again. Somewhat paradoxically, Andrew concluded the interview by reflecting on the importance of 'fighting back' against doxxing:

> When I hear people talking about doxxing [in the US], I try to give them advice that you can't run away from your fears. You have to stay, face your face and just be you, don't go anywhere. You trying to run away from your fears, it's going to make your fear chase you harder … they should try to fight back, try to look for a solution to solve their problem. (Andrew)

Resources for 'fighting back' and being 'strong' by receiving support from their families and communities were reported by most male participants throughout the interviews. Relations of care through family bonds illustrate the alternative of what can be achieved when family social life is a source of strength, rather than a source of shame. Joe reflected on positive relationships, such as his relationship with his uncle, who modelled confidence to him during his experience of being bullied at school after being doxxed:

> So I got encouraged, I was boosted by the confidence I had around me. I like my uncle, he was always calling me often to try to tell me I shouldn't worry. (Joe)

Critical positive masculinity (Lomas, 2013: 183) encourages the 'potential for men to lead happier, healthier, more connected and fulfilling lives', particularly through noting the importance of positive relational bonds. This is also supported by inclusive masculinity scholars, who document 'a widespread rejection of the homophobia that was once considered central to masculinity, alongside an increasing physical and emotional closeness between male friends and, importantly, teammates in high school, collegiate, and professional-level sporting teams' (Burrell, Ruxton, and Westmarland, 2019; Ralph and Roberts, 2020: 20). For Joe, the affirming relationship with his uncle was a tangible strategy and benefit in managing the bullying and harassment spurred on by his doxxing. When asked how doxxing had impacted his life, Joe noted that 'some people commit suicide because of these things … I wonder what I would've done if I had not been given support by my side'. Affirming support from trusted family members and friends is crucial in helping the victim of doxxing during and after their experience.

However, not all male participants received affirming support from their family members and friends. Instead, they were advised to 'be strong' by people in their support networks, presenting a conditional way for them to express their feelings and needs in the aftermath of being doxxed. Andrew identified that his friends felt 'sorry for him', but encouraged him to 'be strong. Nothing's going to happen, don't worry'. Joe also reported that his father encouraged him to 'be strong'. While these encouragements might seem on the surface to be an example of positive or inclusive masculine friendships, requiring victims to demonstrate strength creates further restrictions for the ways men are able to express their needs and respond to harm. As Kullberg and Skillmark (2017: 270) note, 'men have adopted an idealized cultural image of masculinity that leads to a fear of being weak and showing vulnerability and emotions, traits usually associated with femininity'. This represents a further way that men experience the harmful impacts of patriarchy. By framing care-seeking and accessing professional support as a 'feminine' practice, men become cut off from processes of support and validation of their experiences. As Ozyegin (2018: 24) notes, patriarchy '[devalues] male passion, emotionality, creativity, and authenticity', restricting the ways in which men can express their experiences of harm and their need for support.

This brings a different nuance to this chapter's earlier engagement with honour as a patriarchal dividend. Bringing honour into conversation with strength shows how this standard, when applied to men who experience harm, represses the possibilities for how their victimhood can be expressed. This, in turn, casts a critical eye onto the 'benefit' that honour provides in the first place. Rather than an expansive dividend, honour functions to repress and limit the identities through which masculinity can be performed and recognized. The rupturing of social life through doxxing suggests that honour is difficult to hold on to, maintained at a painstaking cost, and, once lost, is difficult to reclaim without complying with hegemonic norms of masculinity. Critiquing honour is a key way in which we can understand how patriarchal systems of relations encroach on possibilities of affirming, expansive care, and respect for all people.

Conclusion

Chapters 3, 4, and 5 have illustrated how women and men are harmed by patriarchal framings of worth in relationships. As such, the social ruptures of doxxing play out within hierarchies which are constructed and maintained by patriarchal ideals of honour, strength, respectability, and worthiness. There is much to summarize from these findings of the chapters, but three key impacts of the rupturing of social life through doxxing are clear.

Ruptures create 'hierarchies of worth' in relational ontologies of intimate, familial, and interpersonal relationships

Participants in this study reported a variety of familial, intimate, and interpersonal relationships from which they sought and received support in the aftermath of doxxing. However, the relational ontologies of these bonds were not equal in their provision of support and their framing of victims' shame and blameworthiness. As evidenced in Molly's account, mothers become lowered in the hierarchy of worth not only because of the sensitivity of their doxxed information but also in the perceived failure to protect their children from harm. This reflects what Tanzer (1985) identifies as the Madonna/Whore paradox, where mothers are judged against an impossible ideal of nurturing at the behest of their sexual freedom and autonomy. Similarly, daughters become diminished in the hierarchy of worth within their families, where support might be totally withdrawn by parents or siblings, or made conditionally available so long as the patriarch does not find out. Secrecy reaffirms women's status within hierarchies of worth, where patriarchy is the guarantor of honour and women must comply with it. Kandiyoti (1988: 275) notes that 'patriarchal bargains do not merely inform women's rational choices but also shape the more unconscious aspects of their gendered subjectivity'. As shown in Aisha and Tamara's accounts, keeping secrets is akin to keeping the status quo.

Interpersonal relationships among colleagues, school mates, or friends utilize patriarchy as a means of belittling or putting down victims of doxxing. These actions might include selling a friend's intimate photographs and contact information to 'get back' at them, leaking confidential work information to diminish their standing at work, or filming them in compromising situations in order to humiliate them. Both men and women can reinforce patriarchal expressions of harm, just as both men and women can be victimized by patriarchal violence. As Connell (2005b: 298) notes, 'no one is an innocent bystander in this arena of change. We are all engaged in constructing a world of gender relations.' This assists in thinking about doxxing perpetration as a form of harassment that, while gendered, is perpetrated by both men and women. Rather, in thinking about patriarchy as a system in which we are interpolated, playing out gender becomes a process of both compliance and resistance. As Kandiyoti (1988: 277) notes, we either tether ourselves to the 'protection' of patriarchy via submission or radically eschew these paradigms and pave our own way.

Ruptures reinforce the harm of patriarchy for all people, but with different gendered outcomes

While it was not the focus or intention of this chapter to create a hierarchy of 'seriousness' or other judgements about the type of doxxing experienced, the patriarchal underpinnings of non-consensual disclosure have shown that doxxing ruptures social life *along gendered lines of distinction*. As noted in the introduction of this chapter, Braidotti's (2019c: 36) accounting for time acknowledges that we are always in the process of 'ceasing and becoming' who we are. By considering the timelines through which doxxing harms are experienced, Chapter 4 noted that the women in this study experienced doxxing as a device of shame which came outwardly from their social worlds and from their own internal self-judgement. For Odette and Tamara, the ways in which they were shamed by their communities extended long beyond the initial doxxing of their intimate photos and videos, interning them in a cycle of shame from which they felt they could not escape. The ancillary harms of second-order doxxing, as reported by Molly and Aisha in their fear of future doxxing attacks, indicate that the rupturing of social bonds is slow to be sutured. For the women in this study, the rupture of doxxing 'ceased' their ability to participate freely in their close and intimate relationships, in addition to their communities. These ruptures also marked a new process of 'becoming' – a starting point of recovering and making efforts to reclaim their autonomy over their information, although not without great personal cost (Braidotti, 2019c: 36).

The rupturing impact of doxxing for the men of this study, contrastingly, appeared to play out along different timelines. While the men of this study overwhelmingly experienced doxxing as a form of humiliation, having prior access to the dividend of honour gave them different strategies to reclaim or restore this status. For Joe, Walter, Andrew, and David, receiving support from their immediate families validated that the honour lost through doxxing was 'rightfully' theirs to reclaim through the performance of 'overcoming' hardship through 'strength'. Materially, the men of this study had significant access to support via their families, including seeking counselling (Joe, Walter, and David), reporting their doxxing to authorities (David), and even moving overseas (Andrew). As noted earlier in the chapter, these offerings of support dramatically contrast to those available to the women of this study. Patriarchy, therefore, underscores not only the different *types* of doxxing levelled against men and women but also influences the *timeline* of harm on a different axes of gender identity. Men experienced doxxing as a harm that they 'overcame' through 'strength' and with family support, whereas women experienced doxxing as a harm that stuck with them 'forever.'

Ruptures present a roadblock to healing, justice seeking and recovery post-doxxing, but strategies of resistance are possible

The harms of doxxing play out over time, considering how virality and the affordances of networked connectivity play out to produce ancillary and second-order harms. By critiquing how patriarchy underscores social relationships, an attunement of the different timelines of recovery post-doxxing has also emerged. At times, these critiques can skew into pessimism – do the gendered differences of doxxing condemn women to live within cycles of shame and blame, whereas men are afforded different strategies to 'overcome' and move on from their experience of harm? A variety of experiences of struggle, overcoming, and healing were reported by participants of this study, regardless of their gender. Rather than lean into pessimism, Chapter 6 sketches out how victims of doxxing can take up critiques of patriarchy to *produce* resistance in the aftermath of non-consensual disclosure.

Ruptures connect with gendered expressions of harm precisely because of this framing of time. Connell (2018: 342) describes this as the 'ontoformativity of gender – the way gender practices bring social reality into existence though time, transforming their starting-points in the historical process that we call society'. This is helpful in thinking about the way that time figures into the unique and gendered experiences of doxxing. Time produces social realities – here, the reality of ruptures – into different configurations of 'starting points'. I take up the related idea that this might also apply to 'endpoints'. Where is the end timeline of a man's experience of social humiliation vis-à-vis doxxing? How might this compare to a woman's cyclical and ongoing second-order harms and threats to safety resulting from being doxxed? Framing timelines of gendered difference is not to condemn male responses to doxxing nor to overstate the challenges faced by women in responding to doxxing. By noting the differences of experience, timelines, and possible support after doxxing, gendered differences of harm explicate new strategies through which victim–survivors can remake their informational autonomy.

I noted earlier in Chapter 4 that conceptualizing the violence of doxxing as social rupture can give the appearance of pessimism but also presents tools and possibilities for a *way back* from the violence of non-consensual disclosure. Just as doxxing harms coalesce in the social, so might social bonds of care present opportunities to reclaim, resist, and amend the harms of doxxing. The final findings chapter explores these strategies of reclaiming and resistance in further detail. But to conclude this chapter's engagement with the concept of social rupture, I draw on Lindemann's (2010) work with men experiencing disability. In his reflection on the 'leakyness' of masculinity gone 'awry' through injury, disability, or accident, Lindemann (2010: 37) notes that 'some gaps can't be plugged, some leaks can't be contained'. Borrowing from this term, the ruptures emerging after doxxing might be conceptualized as a gap or leak that cannot be contained. Information disclosures have been termed 'leaks' well before the internet, harkening back to whistleblowing and spy-craft (Bellaby, 2023). This chapter has considered how the harm of doxxing might be experienced as dread, frustration, or fear of the unplugged leak. The following chapter will consider the options for salvaging, suturing, or stitching up these ruptures as a form of resistance.

6

Resistance in the Aftermath of Doxxing

Introduction

Reclaiming autonomy over personal, identifying, and sensitive information in the aftermath of doxxing might seem like an impossible feat. Once doxxed, personal information is at risk of being disclosed without consent in perpetuity, sparking possibilities of second-order harms from the initial disclosure (Anderson and Wood, 2022). As Wood et al (2025: 14) highlight, 'we live within a system of techno-social interdependencies that confer benefits on individuals who are able and willing to use particular technologies and harm those who are not'. The second-order harms of doxxing, which rupture autonomy over information and impede the freedom of victim–survivors to participate in digital social life, mark the complicity of techno-social spaces with the fraught protection of privacy and data rights.

While the risk of second-order doxxing is serious, victim–survivors are not without strategies to resist the harms of non-consensual disclosure. As I discuss in this chapter, they do so by forging strategies of 'scrappiness', pulling away from locations of harm, and reclaiming their autonomy over their information. In this way, the processes of remaking and rebuilding in the aftermath of doxxing represent new ways of reinvigorating a sense of identity as an informational subject by renewing the boundaries of where information is shared, and to whom. Uldbjerg (2021a: 28) has challenged digital abuse scholarship to resist against the narrative of the 'ruined' victim, who cannot move beyond their experience of harm. This chapter responds to this challenge by taking inspiration from the strategies evidenced by the participants of this study as a way of understanding the resistance that can emerge post-doxxing.

Key to understanding the capacity to resist, reclaim, and remake oneself in the aftermath of doxxing is confronting the question of informational autonomy. Elsewhere in this book, I have discussed how autonomy over information draws upon Nissenbaum's (2018) concept of contextual integrity, where the ability to control how information is disclosed, to whom, and in what contexts is framed as the guarantor of privacy. As Eckert and Metzger-Riftkin (2020) have reflected, the bounds of contextual integrity play out along gendered lines, as the discussion of violence in Chapter 3 explored in further detail. However, this chapter departs from these perspectives by considering the extent to which we truly possess autonomy over our information.

Our information has never been more widely accessible, with personal data, info-metrics about our consumption, entertainment, and behaviours captured in an ongoing 'politics of knowledge in the digital age' (Zuboff, 2022: 3). Autonomy over information appears especially tenuous when we consider the many ways that our images, movements, behaviours, choices, and even our desires are tracked, logged, surveilled, and anticipated by modern digital technologies (see Elias, 2024). This may provoke a feeling of despondency that non-consensual disclosure is inevitable due to the lack of data protections online. But to confront the harm of doxxing is to situate these disclosures in the context of ever moving, shared data *of* ourselves. With this in mind, this chapter considers the alternative of *informational autonomy*, which affirms the right of all people, but especially victim–survivors of doxxing, to enjoy full control over how, when, and to whom their information becomes known. This autonomy over information plays out in subtle, contingent, and minor ways. The participants of this study highlighted myriad ways that their informational agencies were sparked, renegotiated, or animated in the aftermath of doxxing.

In this chapter, I argue that informational autonomy provides victim–survivors of doxxing with renewed options to navigate their digital lives on their terms in the aftermath of non-consensual disclosure. Piecing agency over information back together is often 'messy' work, and as Lavorgna, Ugwudike, and Sanchez-Benitez (2022: 253) highlight, can spill into forms of resistance which form new 'secondary harms', such as counter-disclosures or erasures of personal access to digital spaces. Regardless of its form, reclaiming informational autonomy is messy work which 'attunes us to the generative potential' of finding new strategies of resilience in the aftermath of harm (Anderson, Lundberg, and O'Donnell, 2024: 2). Noting the resilience of victim–survivors in the aftermath of harm should not imply that they bear the responsibility to 'learn to cope better' (Green, Calverley, and O'Leary, 2021: 857). Rather, adopting a 'strength-growth-resilience' framework to victim–survivor responses draws our attention to 'how individuals and communities generate their sense of identity through the inclusion of some, and the exclusion of other, experiences of victimization: the battles they fight; the victories and defeats they experience; the ways in which old relationships are broken and new ones formed' (Green, Calverley, and

O'Leary, 2021: 857). In this framing, victim–survivor strength and resilience sits alongside loss, change, and breaking points, not as a contradiction, but as a complex accounting of the continuum of victim experiences.

I do not set out a uniform or generalized consensus of what 'counts' as autonomy, agency, and self-determination, but note that resisting and reclaiming in the aftermath of doxxing entails a process of scoping out 'what feels right' to the individual. Victim–survivors of doxxing have extremely limited options to seek justice through the criminal justice system. As noted in Chapter 4, doxxing is not criminalized consistently and where they exist, legal definitions lack clarity. None of the participants in this study who reported their doxxing to police were successful in bringing a complaint under a criminal or civil statute, with many participants unaware of their legal rights. For many victim–survivors of doxxing, seeking justice after doxxing cannot be facilitated within the criminal justice system.

As such, this chapter draws upon the work of Fileborn (2014, 2017), O'Neill (2022, 2024), Henry, Flynn, and Powell (2015), and Loney-Howes (2018, 2020) on online informal justice as an alternative to formal justice seeking. Strategies of online informal justice, including counterpublics, digilantism, and forms of narrative victimology, centre victim–survivor voices and stories in an effort to 'return the prospect of ownership to victims which criminal justice agencies have historically attempted to retain for themselves' (Cook and Walklate, 2019: 251). Online informal justice foregrounds the importance of collective action, voice, and resistance outside of the criminal justice system. This chapter takes inspiration from these framings to consider different forms of victim agency that exist beyond the criminal justice system.

Using resistance as a conceptual tool in this chapter is not to promote myths of 'emerging stronger than ever' in the aftermath of harm. The previous chapter critiqued similar constructions of harm as the pigeon-holing of masculine understandings of harm through 'strength', 'honour', and a requirement to 'overcome'. Green, Calverley, and O'Leary (2021: 854) also stress the importance of not 'assuming victimization will always lead to positive personal growth' when accounting for the resilience of victim–survivors. Resilience and resistance, therefore, are not expectations of victim–survivor responses in the aftermath of harm – and therefore, a judgement of the 'good' or 'bad' victim – but a way of drawing inspiration from the ingenuity, scrappiness, and pragmatism that victim–survivors already possess within themselves. From a feminist empiricism standpoint, there is much to learn from victim–survivors by paying attention to the strategies and tools that are deemed useful and meaningful to them, rather than projecting an academic judgement about 'what counts' as justice seeking, sense making or recovery after harm.

Withdrawal

Withdrawal from digital and social spaces was a key strategy which the participants of this study used to resist against the harm of doxxing and its repercussions. Spaces to withdraw from and to were represented in a range of ways across participant reflections. Experiences of digital harm can be difficult to process as the spaces in which these harms occur are often hard to disengage from (Kuntsman and Miyake, 2022). However, many participants demonstrated a 'scrappiness' in the ways that they counteracted and responded to ancillary harms of doxxing, such as harassment and bullying, by withdrawing from social environments where they might be at risk of ongoing victimization. 'Scrappiness', as Ybarra-Frausto (1989: 7) explains, is to be 'proudly resourceful', finding solutions by embracing the 'inventiveness at the heart of surviving oppressive material conditions' (Ybarra-Frausto, cited in Monreal, 2019: 120). The scrappy inventiveness of 'misusing' technology is one such way that victim–survivors used withdrawal to disrupt further harassment, such as those demonstrated by Tamara and Jessie in the following excerpts:

> I think I just had to accept it the way it is. Cause a lot of questions were coming. And I think it went viral. It was mostly from people … who knew [doxxer and I] were friends. So I just left most of the messages on read, I didn't want to answer to any of them or give more explanation or some answers to their questions. (Tamara)

> I don't think I was in a good position and the right state of mind to explain to people what actually happened. 'When you have seen it, just take it away.' That's all I said to myself. So whatever you've seen out there, just take it. I understand that trying to explain to people could cause me more harm. Cause no one would understand that. (Jessie)

Opting out and leaving harassing messages 'on read' were strategies of digital disengagement Tamara and Jessie used to filter out the second-order harassment that followed their doxxing disclosures. Digital disengagement, a term increasingly popular in describing ways to manage digital technology fatigue, is evident here in Tamara's negotiation of how she engaged with technology to mitigate the flow on effects of her doxxing. Kuntsman and Miyake (2022: 4) note that the 'forced incorporation' of digital networks, social media, and other technologies which facilitate social connection can be a conduit for harm by sustaining a 'compulsory digitality and connectivity'. Digital disengagement, as a strategy of rejecting 'compulsory digitality', is an elastic process where actors may be able to disengage in some contexts but not in others. This is an example of a temporary, faltering resistance to the ancillary harms resulting from doxxing.

Later in this chapter, I explore the stop-start nature of autonomy in response to doxxing as 'faltering', but here I note that time both underscores the nature of cyclical doxxing harms and is also disrupted by acts of resistance, such as withdrawal. Time is present in Tamara's account as an accelerant of the ancillary harm of doxxing. She was well aware that questions, intrusions, and further harassment resulting from being doxxed were 'coming', and perhaps would continue to come. Opting out of engaging with harassment by refusing to clarify the context of her doxxing is a temporary way of stemming the tide of ancillary harms. Further, as Ahmed (2014: 201) suggests, 'finding a more liveable way to sustain silence' through withdrawal affirms the victim–survivor's agency to determine in what environments they participate, with whom and when.

Jessie also invokes a timeline of harm after doxxing where she makes efforts to minimize how and to whom she remained accessible to. Jessie was doxxed when her Facebook account was hacked and intimate photos were posted, an experience she described as 'difficult to try to convince people that I wasn't the one [making the posts]'. After her initial efforts to regain control over her account, and to convince her family and friends that she was not behind the posts, Jessie embodied strategies of opting out by refusing to explain. In Chapters 4 and 5, distinctions between supported, believed, and credible victims were highlighted as being contingent upon gendered stratifications of 'worth'. For women like Jessie, being doxxed through sensitive, personal, and intimate images and videos of their bodies interpolates them to the bottom of the hierarchy of worth. Jessie sought support which was denied by those who believed her to be lying, or to be deserving of harassment due to underlying patriarchal judgements about women's sexuality and shame. What she did instead, however, constitutes a form of resistance because she affirmed for herself that her 'own definition of what counts' was more important to her healing than explaining or justifying herself to others (Walklate, 2014: 74). By '[understanding] that trying to explain to people could cause me more harm', Jessie opted out of performing contrition or other forms of shame expected of an 'ideal' victim, instead choosing a course of action which protected her own interests and safety.

Withdrawal as a demonstration of agency was also evident in how Hara explained her need for solitude in the aftermath of being doxxed:

> So I think what really helped me was that I just stayed by myself. Doing things alone. Because personally, if I'm going through stuff like that or I'm down, I prefer to stay by myself, just reflect on what is going on without talking to people. I think I feel better that way. (Hara)

Withdrawal played out across a variety of social environments and contexts. In Hara's case, withdrawal from her social life resulted in solitude and physical withdrawal from environments and spaces where she might encounter others. Withdrawal carved out a space for Hara to reflect on her experience and negotiate how she wanted to engage with her community on her terms. It should be noted that not all victim–survivors of doxxing would frame time away from their communities and social bonds as desirable; other participants noted that they felt ostracized by their families and friends in the aftermath of doxxing (see discussion of Molly and Aisha's experience in Chapter 4). What is critical to distinguish between these experiences is the agency of Hara's *choice* to withdraw on her own terms.

Taking time to reflect in the aftermath of doxxing also affords victim–survivors strategies to engage with justice-seeking on their own terms. For those who choose to pursue a criminal justice remedy, victims are often required to fit within the timeframe of reporting requirements, court procedures, and the challenge of holding the timeline of events in place to meet evidentiary standards (Clark, 2010; Bluett-Boyd and Fileborn, 2014). Clark (2010: 34) notes that these evidentiary standards, such as '[responding] to a set of yes/no questions that break down any personal attempt to construct a coherent and meaningful narrative', deny victim–survivors the opportunity to frame their experience in their own words and voice. In particular, the evidentiary burdens placed on victims during trials require them to rehash trauma while also keeping the details and timeline of their testimony in order. Brison (2002: 108) describes her own experiences as a victim–survivor in court, describing the relief she felt after the verdict was passed down as an 'unclenching': 'I didn't have to keep the story straight. It didn't matter any longer if I got it right. I could let go of the details I'd kept alive in my mind, the narrative I'd remembered, rehearsed, and, finally, delivered to the court.' While much research into informal justice has highlighted the importance of community building and sharing stories as strategies of resistance (Fawcett and Shrestha, 2016; O'Neill, 2018), Hara's account shows how withdrawal is a form of agency – and therefore, a strategy through which victim–survivors might seek justice on their own terms.

Coupled with strategies of withdrawal and leveraging silence, Hage's (2009: 76) notion of 'endurance' is also key to understanding how victim–survivors might opt in and out of what is expected of them in the aftermath of harm. Hage (2009: 76) considers endurance to be 'asserting some agency over the very fact that one has no agency by not succumbing and becoming a mere victim and an object in circumstances that are conspiring to make a total agentless victim and object out of you'. This is a very helpful way of thinking through the strategies of renegotiating what information is available, visible, and knowable about the self in the aftermath of doxxing, particularly when the doxxed information lends itself so well to objectification. In Chapter 3, Odette described her image-based abuse doxxing as painting her to be the 'cautionary tale' to the daughters, wives, and other people in her community. But she also modelled endurance by refusing to explain and by standing aloof to the requirements of the 'ruined victim' to perform contrition (Uldbjerg, 2021a: 28).

Similarly, Tegan's refusal to deny or clarify her friend's disclosure about her abortion can be understood as a form of endurance. By refusing to engage with her doxxer – by 'letting her [post] whatever she wanted to' – Tegan not only endured the harm of non-consensual disclosure but gave no rise or fodder to fuel further harassment or abuse. Tegan demonstrated how leveraging silence helped her to resist against the ancillary harms of doxxing, such as bullying, harassment, and social fall-out. Using silence also became a strategy of reclaiming her control over her information by taking back the choice to disclose, affirm, or comment on her information on her own terms.

Confrontation and forgiveness

For many victim–survivors of doxxing, confronting their perpetrator is neither desirable nor possible. However, when the perpetrator is known to them, confrontation may represent a strategy of resistance for some victim–survivors. In this study, confrontation was imagined and performed by participants with varying motivations: to understand the rationale or motive of their doxxer; to seek recognition for the harm suffered; and to be given an apology and an opportunity to forgive. As Sabiyah reflects, confronting her doxxer – her ex-partner who shared her intimate photos after they ended their relationship – was a strategy not only to resist against the harm of doxxing but was also a way of gaining back control over her information.

> I called him. I was very angry. I talked to him, why would he do such a thing? He actually apologized that he was so sorry, he was wrong and deleted everything. (Sabiyah)

Through confrontation, Sabiyah was able to seek recognition and accountability from her perpetrator. McGlynn and Westmarland (2019: 188) note that 'recognition is more than "being believed"', but also 'encompasses the significance of the experience being acknowledged'. While many victim–survivors may not be able to gain recognition from their perpetrator that their actions were 'wrong', Sabiyah's account shows how, in some cases, confrontation can set in motion pathways to reconciliation and forgiveness. Anderson (2016: 116) notes that prior to forgiveness, anger plays an important role in validating the victim–survivor's experience of abuse as 'anger … begins the process of recognizing the injustice for what it is'. This framing is echoed in Ahmed's (2014: 201) work on the politics of emotion, where 'anger against injustice can move subjects into a different relation to the world'. Anger, in Sabiyah's account, was not only a catalyst for confrontation but a means of repositioning Sabiyah in relation to her perpetrator. In other words, anger was one way that she could 'act back' in the aftermath of doxxing. This also shows the need to hold space for the contradiction of being angry while also being open to hearing and accepting an apology. After the apology, Sabiyah 'did not think of reporting' her perpetrator to the police, as she was satisfied that he had deleted the post of her intimate photos. Embracing a continuum of victim–survivor agency calls for '[responding] to injustice in a way that shows rather than erases the complexity of the relation between violence, power and emotion' (Ahmed, 2014: 196). As such, anger holds just as much weight in being the catalyst for victim–survivor resistance as the seeking of healing or redress.

Confronting perpetrators in the aftermath of doxxing brought up a range of other emotional responses for participants, including closure and acceptance. For Joe, confronting his doxxer did not provide him with answers as to the motivations behind his doxxing, but opened up a different pathway to compassion and understanding.

> I confronted my brother's friend that actually got the video … So I told the boy what happened? Why did you get the video? Yeah, he had no answers. For me, I just try to understand that these things, he might be ignorant of whatever he might be doing or stuff like that. Yeah, I, I just forget about most of these things because I know people who think people will try to do whatever they need to do. (Joe)

Here, Joe showed an internal sense-making of his experience by understanding that, just like him, his perpetrator was a young man prone to making mistakes. This form of confrontation led to efforts to mutually understand each other, outside of the dynamics of victim and perpetrator. Joe was able to 'make himself available' for reconciliation, a strategy which Eisenman (2004: 57) identifies as key to accessing the structural privileges of accountability within male peer relationships. Accountability, in Joe's case, provided for strategies of healing through confrontation as his perpetrator was willing to engage in dialogue. In contrast to other victim–survivors in this study, Joe's relation to his perpetrator played out along inclusive and positive masculine relations, where rapport was made possible through mutual (and hegemonic) assurances (see Ralph and Roberts, 2020).

Positive masculinity scholars suggest that 'interactions between friends are part of gender performances', where male friendships become iterations of social norms and codes of 'doing gender' (Migliaccio, 2010: 226, 229). As discussed in Chapter 5, these norms can lead to reductive requirements that men 'overcome' harm through performances of 'strength'. However, what is apparent in this account is that the pre-existing relationship between Joe and his perpetrator offered them a frame of reference that they could return to – one that allowed Joe to interpret his doxxer's actions as misguided but not malicious. Joe's use of confrontation provided him a form of accountability and closure which, while possibly inaccessible to the women of this study, afforded him a way of resisting against the harms of non-consensual disclosure by healing his relationship with his perpetrator.

In these accounts of confrontation and forgiveness, there is a need to make a critical distinction between victim–survivors *choosing* to forgive and move forward on their terms and feeling *pressured* to forgive as a way of resolving their experience. Restorative justice scholars have critiqued the pressures placed on victim–survivors to forgive their perpetrators in justice conferencing and mediation settings, noting that apology-seeking and forgiveness often play out in gendered ways, where female victim–survivors may feel pressured to 'grant' forgiveness in order to bring justice to a close (Stubbs, 2007; Daly and Stubbs, 2017). However, promoting a victim-centred approach to justice is to centre victim–survivor desires, needs, and articulations about what justice 'looks' like to them. Sabiyah reports that she found that justice could be accessed through forgiveness and accountability outside of the criminal justice system:

> I didn't wanna involve the police. Because when I went to him to talk to him about the photo, he apologized, so I feel no need to involving the police … seeking an apology from the person who's wronged you can be really healing. (Sabiyah)

Sabiyah's framing of healing in this excerpt resists against criminal justice imaginings of victim redress, showing instead an example of victim-initiated justice through relationality. This is an example of resisting against the criminal justice system and its individualistic framings of victim–survivor and perpetrator relationships. As Haraway (1991: 135) notes in her critique of western understandings of agency as 'ownership of the self', where a subject '[holds] a core identity as if it were a possession', we have much to gain by embracing relational and collective outputs in place of individualistic framings of justice. Resisting against carceral imaginations of justice, which might include arrest, trial, and punishment of a perpetrator, in favour of an interpersonal healing through apology and forgiveness is one such way that justice can be imagined as relational. This framing suggests that agency is experiential, with the subject iterating their agency through repeated efforts and endeavours, as Sabiyah embodies in her framing of seeking and receiving an apology as 'healing'. What this might mean for victim–survivors of doxxing is that agency in the aftermath of non-consensual disclosure does not have to look like retrieving or reclaiming their information as an act of repossession. Instead, agency can play out in myriad ways that affirm their informational selfhood *iteratively*.

Iterative examples of agency were also shown through participants' willingness to share their experiences, foster connection, and build community. Victim–survivors of doxxing cultivate 'just spaces', regaining 'a sense of belonging' by 'exercising [their] right to everyday life whereby [they] make space [their] own by moving through it as a part of [their] embodied everyday practices' (Antonsdóttir, 2020: 739). As Eden reflects, sharing insights and reflections in the aftermath of being doxxed is one way of cultivating this 'sense of belonging':

> Keeping it back to myself will not help anybody out there. But sharing it out will actually help one or two persons have an idea, okay, this is life, or this is the approach I should take in order for me to better in my life. So it's something I can easily, happily share out in order to help people not to make that same mistake. (Eden)

Eden spoke at length throughout the interview about the sense of pride and meaning that she found in sharing her experience of doxxing with other people in her community. Additionally, she noted that participating in the study was something she was proud to do, with the hope that her experience could reach others who might be facing similar harms of non-consensual disclosure. This reflects the emancipatory potential of what Heemsbergen (2016: 144) terms the act of 'voluntary disclosure' of personal information. These motivations also reflect a narrative victimology understanding of victim–survivor agency, where interviews and dialogue 'afford victims a space to reassert control and agency over experiences which have been previously disempowering and isolating' (Cook and Walklate, 2019: 241). In contrast to the strategies of withdrawal discussed earlier in this section, Eden's account shows the subjective nature of how victim–survivors decide what is right for them in seeking justice and resisting against harm. Eila modelled a similar motivation to advise and deter people from doxxing, stating that 'when I see people [doxxing] online, I use myself as an example, I just don't put my name out there'. While Eila did not share Eden's willingness to openly disclose her experience, the motivations to intervene in potential moments of harm might be seen as part of 'a process of coming to terms with the impact of the different violence(s) in different lives' (Cook and Walklate, 2019: 250). While being open about past experiences of harm might not be accessible to all victim–survivors, for Eden and Eila, sharing their experiences with others was a key way to seek support and reclaim autonomy over personal information.

Complaint and community building

Not all participants in this study had the ability or resources to speak out about their experience of doxxing to find resolution through apology, forgiveness, or building community. For Chelsea, exercising her voice through complaint was one way in which she reclaimed her autonomy over her information. Complaint brought about strategies to resist the second-order harms of doxxing but also brought with it discomfort and self-doubt about the validity of her experience being harmful 'enough'. Chelsea reflected on her decision to make a formal complaint in the aftermath of her doxxing, where her personal information was leaked in a large-scale telecommunications hacking incident.

> I am actually going to have to fend for myself in a way and actually speak to [telecommunication company] and say this is not okay and that we have a full amount of data leaked and we were part of the same account. So therefore if [my brother's] data was shared, my data was shared and you've confirmed it, why aren't I getting the same amount of support as he is? (Chelsea)

While Chelsea was pessimistic about the outcomes of retrieving her personal information, complaint was an important way of asserting that her information mattered. In Chapters 4 and 5, I explored how the violence of non-consensual disclosure plays out along patriarchal and gendered lines. Chelsea's complaint could also be read through this lens, as her brother, the primary account holder of their family mobile plan, received personalized support after the disclosure, whereas she did not. Ahmed (2021: 298–9, emphasis in original text) suggests that complaints can mark a 'trail' from where violence occurred to a new form of autonomy over the harmful incident: 'a complaint is *what we plant*, a new growth of some kind that marks the site of violence. The site of violence is the site of protesting that violence, saying no to that violence.' For Chelsea, complaining was a vital means of wrangling back control over her personal information, in both applied and symbolic ways. Complaint took the form of her demanding her right to her personal information, and her demands to be taken seriously in finding a resolution.

Speaking out in the aftermath of doxxing took on other forms throughout the participant's reflections. For Eden, speaking out about her experience with doxxing was framed as a liberatory way of connecting to and building community through the sharing of testimony. Bhagawati and Priya (2025: 20) have highlighted the importance of 'radical healing' in communities of victim–survivors, where those who have endured serious forms of gender-based violence (such as acid attacks and sexual harms) can support each other by 'becoming a mirror' to each other in solidarity. Eden shared her experience of being doxxed broadly with her community, suggesting that there are pathways to make new meanings from experiencing non-consensual disclosure:

> It's a way of reaching out to a large number of people, a way of sharing an experience to make them feel better … so it helps you to know,

okay, that you're going through this at this point but it does not mean it's the end of the world. It does not mean that you're stuck in that situation that you will surely pull through. (Eden)

As Eden demonstrates, there is power in the sharing of testimony with others to mitigate shame and centre new bonds of connectedness in the aftermath of harm. This serves as a counterpoint to the earlier chapters which considered the shame embedded in the patriarchal harms of judgement and devaluation spurred on by doxxing. Ahmed's work on collective complaint is useful here in understanding how groups can find solidarity in shared experience, reframing harm on their terms. Ahmed (2021: 6–7) reflects on the purpose of collective complaint, where 'their collective became ours. I think of that ours as the promise of feminism, ours not as a possession but as an invitation, an opening, a combining of forces'. Eden discussed her desire to empower other people, particularly young girls, by sharing her experience of doxxing without shame. This also echoes what O'Neill (2022: 568) discusses as the importance of speaking out but also 'speaking in', finding resolve and affirmation through collective narrative and storytelling in trusted and safe communities. While Eden's account describes a strong community where she gave and received support, it is also important to note that not all victim–survivors have access to communities who are willing to listen, nor does it guarantee that their grievances will be collectively heard in the spirit that the victim–survivor desires (Fileborn, 2014; O'Neill, 2022).

Eden's reflections on the affirming relationships that can emerge after doxxing can be read through a Floridian understanding of the informational self, extending understandings of how autonomy can be performed and experienced online. Floridi (2011: 565) suggests that through 'tinkering with the self', subjects experience both expansive and inhibiting forms of selfhood in the onlife. He considers that information–computer technologies are an 'egopoietic technology' of self-construction that 'significantly [affect] who we are, who we think we are, who we might become, and who we think we might become' (Floridi, 2011: 551). Eden's use of technology to create community and support in the aftermath of being doxxed is one such example of the 'egopoietic' potential of using technology to craft responses in the aftermath of harm. This also highlights how informational self-determination, another Floridian conceptualization of 'tinkering with the self', might present new negotiations for privacy, access, and visibility in digital spaces. Informational self-determination has a rich history in data sovereignty literatures (see Walter et al, 2021; Walter and Suina, 2023) as well as governance practices and regulations for data collection and extraction (see Fordyce and Jethani, 2021). Coupled with the insights and experiences of the participants of this study, the last section will explore these literatures to think through affirmative models of informational autonomy through which we might practice, tinker, and live out self-determination over our private and personal information.

While it is important to acknowledge and validate the desires of some victim–survivors to speak out and share their experiences, it is also important to note that these forms of resistance may not be accessible to all victim–survivors of doxxing. Autonomy over information is not conceptualized in static terms throughout participant experiences in this study. Indeed, many participants, such as Jana, reported a journey of faltering autonomy, where their efforts to regain informational autonomy went through varying attempts and levels of success. Validating the unique forms of victim–survivor autonomy in the aftermath of doxxing requires moving beyond judgements of outcomes and success, but instead seeing it as an active process of trying and trying again.

In Jana's experience of doxxing, her efforts to regain informational autonomy oscillated as she pursued strategies of resistance across different aspects of her life. Jana was doxxed by a coworker who made threats toward her safety and shared her home address online. Jana's workplace took her complaint against her perpetrator seriously, which resulted in him being fired. Additionally, as much of the doxxing had taken place on X, Jana reported him in breach of the Community Guidelines and was successful in having his account suspended. While Jana was still very concerned about her physical safety – as he had threatened to come to her home – she secured some substantial protections from him in her workplace and in her digital space, illustrating how the affordances and design of digital architecture can be mobilized to protect victim–survivors from the enduring impacts of doxxing. Successfully using the Community Guidelines policy to block and suspend her abuser's account can be interpreted as an 'egopoietic' use of technology to create a digital space where she was free from his ongoing abuse and targeted harassment (Floridi, 2011: 511).

In contrast to other participants' experiences of seeking justice, it may appear that Jana was able to find justice outcomes which addressed her immediate proximity to her doxxer. However, Jana's ongoing anxiety and concern for her well-being illustrates the imbrication of the harms of her doxxing, as she feared that the doxxed information could resurface and reignite the cycle of harm for her later in the future (Anderson and Wood, 2022: 197). Jana's strategies of coping, making do, and 'getting on with it' are key examples of how autonomy over information can falter, or be successful in some instances but imperfect in others. Rather than frame faltering autonomy as a weakness, the incremental efforts which comprise justice seeking through 'small, cumulative, and interconnected events and responses' help to sketch out an iterative framing of justice seeking and informational autonomy as intertwined (McGlynn and Westmarland, 2019: 196). This also illustrates the 'kaleidoscopic' nature of justice, where 'justice is not linear, but has multiple beginnings and possible endings', reflecting the unique and individual needs of each victim–survivor (McGlynn and Westmarland, 2019: 186).

Articulating the informational self

In Chapter 5, I critiqued how patriarchal honour systems construct 'strength' and the ability to 'overcome' as requisites for achieving justice in the aftermath of doxxing. While these value judgements can be limiting frameworks to understand the harm of doxxing when they are forced upon victim–survivors, it is also important to note that strength and resilience can be important ways that victim–survivors self-interpret their responses to non-consensual disclosure. This section will consider how individual conceptualizations of strength and resilience help us to understand a range of victim–survivor framings of resilience, without promoting these responses as required or indicative of the experiences of all victim–survivors of doxxing.

For Tamara, her ability to endure and fight on in the aftermath of doxxing required mobilization against the apathy and ineffectiveness of the traditional justice system. Tamara made efforts to report her doxxing to the police, who took her statement and interviewed her doxxer but did not keep Tamara up to date with the progress of her complaint. In the following excerpt, Tamara's frustration with law enforcement's unwillingness to take her doxxing seriously and to support her is evident:

> I just feel like the system should improve and do more, do better things. Especially look after this person who was being doxxed. I was just being very strong. Another person could be afraid to and not be very strong like me. Another person might commit suicide due to everything that surrounds your personal details going viral. Not everybody might be strong like me. (Tamara)

Unlike the previous framings explored in Chapter 5 of strength being intertwined with masculinity, Tamara frames her strength and resilience in *spite* of apathetic responses from the justice system. Ahmed (2014: 197) cautions against narratives of strength, as 'overcoming bad feeling is not a measure of healing. Indeed, if "bad feeling" is also an effect of injustice, then to overcome bad feeling can also be to erase the signs of injustice.' Tamara is not touting that her strength and resilience is the only way forward, but notes that many other victims of doxxing might not have the resources or capacity to be strong *in spite* of the criminal justice system. Tamara's experience of ineffective support from police is unfortunately not uncommon, reflecting other victim–survivor experiences of fearing not being believed, or feeling uncertain about how to meet evidentiary burdens when reporting digital abuse to police. Uldbjerg (2023: 536) critiques legislative responses which treat forms of digital abuse, such as image-based sexual abuse, as a youth issue, noting that this can silo adults who experience this form of harm. However, Tamara's refusal to gloss over her 'bad feeling' is also a way that she demonstrates her autonomy. The 'bad feeling' of anger and injustice about her treatment by police has instead been mobilized as a form of critique, and a demand for better support for other victim–survivors in the future.

Frustration with the apathy of systemic processes in the aftermath of doxxing extend beyond critiquing the criminal justice system. As Chelsea notes, corporations who are negligent about the personal information of their customers do not typically take responsibility for leaks, leaving much of the responsibility to mitigate the risk of doxxed information to the victims:

> I don't think I'll ever deny myself the ability to feel sad about it. I think there's part of me that goes, well what happened, happened, it is what it is. It's out of my hands now and there's no, there's no responsibility other than to look after myself. I can't be responsible for how someone else does it. And the other half of me just goes, I feel like a victim of someone else's problem. I feel like I've been indirectly affected because someone didn't cross their Ts or dot their Is. (Chelsea)

Chelsea utilized complaint as a means of exercising her agency in the aftermath of doxxing. Here, she evidences the difficulty of wrangling back her informational autonomy as an issue of personal responsibility. After her information was leaked by the telecommunication company, Chelsea engaged in an arduous and drawn-out process to close her account and move to another provider. This process was significantly disrupted by the structure of her account falling under a family plan in her brother's name. Despite her brother also closing the account and leaving the company, Chelsea was stuck in a state of limbo where she did not have the 'authority' to make the decision to close her account without her brother's sign off, despite the account ceasing to exist. Chelsea's frustration with this drawn-out process illustrates what Kuntsman and Miyake (2022: 26) describe as the 'elastic' possibilities of digital disengagement, where 'spaces, times and practices of opt-out, and the possibilities of digital disengagement, open and close based on an unequal distribution of economic, socio-cultural and digital capital'. In Chelsea's case, her ability to 'opt-out' and digitally disengage from a corporation who had harmed her and put her at risk by sharing her personal information was 'closed off'. This is not to suggest that some victim–survivors of doxxing are better equipped to totally disengage or reconcile the harm that non-disclosure creates. However, it is important to note, particularly in terms of economic resources, that some victim–survivors have access to possibilities of disengagement that others do not. Chelsea's account shows how valorizing the reclaiming of doxxed information is implausible, but that victim–survivors negotiate what outcomes matter to them *in spite* of systemic apathy and indifference to their needs.

Self-determination as informational subjects

In the previous sections of this chapter, voice has been evident in a variety of ways: through withdrawal, in confrontation, by sharing stories, and issuing complaints. These forms of voice are unique, at times contradictory, and differ depending on the individual justice needs of the victim–survivor. In thinking through what it means to tell your story, narrative victimology has shifted focus from 'the measurable nature and extent of violence [to] how violence is experienced and lived from the point of view of the victim' (Cook and Walklate, 2019: 241). To share details of your lived experience, to give advice to others, and to have experiences validated is a form of resistance made possible through victims' agency to share their stories in their own words and to practice self-determination in doing so. While much of the literature surrounding informational self-determination is grounded in legalistic and rights-oriented understandings of personal data, the final section of this chapter will consider how the voices of victim–survivors *expand* possibilities of resistance. Following on from Floridi's (2011: 565) concept of 'tinkering with the self', this section outlines how victim–survivors use their voices to work through and on the site of harm, in addition to how voice lends itself to creativity as a strategy of self-determination. These examples of victim–survivor voices illustrate that digital harm scholarship can move away from 'ruination' framings, and instead account for the creative and crafty counter-responses that are possible through self-determination (Uldbjerg, 2021a).

Informational self-determination has been addressed as a legalistic framing of data rights, primarily as the right to be forgotten (see Corbridge, 2018). Elsewhere, Fordyce and Jethani (2021: 775) discuss a related concept of data provenance, where locating the place of origin of data therefore situates it within stronger associations of ownership, control, and consent prior to its use. These framings are a starting point in scoping out informational self-determination after non-consensual disclosure. However, this chapter contributes to these framings by considering what might be *expansive* in the expression of informational self-determination, not only restrictive. In this vein, informational self-determination can unlock new strategies of resistance in the aftermath of doxxing by framing victim–survivor responses as 'working through' and 'on' their doxxing, rather than 'beyond it'. Brison (2002: 115) speaks about examining relationships with past trauma as learning 'how to relinquish control, to learn by going where we need to go, to replace the clenched, repetitive acting out with the generativity of working through'. Working through is not about the denial of pain, frustration, or anger, but the valuing of those emotional responses as a key beginning place to take action from. Chelsea reflects on her emotional distress and the burden of following up after the disclosure of her personal information.

> Personally for me, when the feelings get too much, it's just something that gets pushed to the side and this is causing me too much emotional pain. So let's just push it to the side. What can I do about it? Not very much. So let's just put it to the side for the moment. Let's work on things that we can fix. (Chelsea)

Chelsea's pragmatism about finding solutions in the aftermath of disclosure does not deny her simultaneous frustration and distress at being put in a situation where she had to regain autonomy over her information. But this also shows that sitting with emotional responses in the aftermath of harm is not extraneous to resisting against harm, but an integral part of it. This form of self-determination blends the seeking of solutions with the validation of emotions stemming from non-consensual disclosure. This simultaneous process of sitting with emotions and taking action reflects what Ahmed (2014: 202) frames as 'working with and on rather than over the wounds' of past harm. Chelsea does not invalidate or stew in her emotional distress but funnels this into action. She does not deny her emotions; she 'puts [them] to the side' while she makes efforts to regain control over her personal information. The act of validating one's own experience could be considered rare in a victimology context, where many victims are made to feel that their experiences are 'illegitimate' and difficult to recognize (Kenney and Craig, 2012; Spencer, 2015). As Chelsea demonstrates, the negotiation of emotion and action is a key way that victim voices perform resistance.

Similarly to Chelsea, Jessie reflected on sitting with her emotions to learn from and continue resisting in the aftermath of non-consensual disclosure.

> I would say I see it as a lesson. To me, I'm determined to [either] learn from that or learn never to trust anyone. I learn to just check things in the world. I also learned that even things you never imagine could happen to you at any time. (Jessie)

Like Chelsea, Jessie's framing does not deny or minimize the extent of harm she suffered as a result of her doxxing. The impact of disclosure has led to a heightened sense of risk, as evidenced by her desire to 'check things' in an effort to anticipate further disclosures of her information. However, Jessie also shows efforts to resist her initial inability to trust those around her. By learning from her experience instead of distrusting those around her, Jessie also shows how self-determination in the aftermath of doxxing requires the validation of emotions. As Ahmed (2014: 202) notes, 'recovering from injustice cannot be about covering over the injuries, which are effects of that injustice; signs of an unjust contact between our bodies and others'. Instead, Jessie strikes a balance between learning from her experience of non-consensual disclosure without it giving way to distrust and isolation from those around her. Further, Jessie does not place limitations on her feelings of anxiety and risk after being doxxed, noting that the risk of non-consensual disclosure can take place 'at any time'. In turn, this pragmatism shows that 'redemption is not a simple and/or measurable moment, but it is a process of coming to terms with the impact of [violence]' and its potential for ongoing impacts (Cook and Walklate, 2019: 250). This is a noteworthy way in which Jessie's voice is centred in her seeking of self-determination; she does not place restrictions or judgements on her journey of sense-making in the aftermath of non-consensual disclosure.

Conclusion

This chapter has reflected on victim–survivor resistance, where autonomy after doxxing represents an iterative, active process of reclaiming control over personal information. The strategies of resistance discussed in this chapter are not singular, but point to the multiplicitous, varied, and, at times, contradictory ways that justice is sought by victim–survivors. McGlynn, Downes, and Westmarland (2017) highlight the importance of recognition, dignity, voice, prevention, and connectedness as key ways that justice can be achieved. In this chapter, I showed that the unique strategies of resistance in the aftermath of harm represent victim–survivor efforts to seek justice on their own terms. This has implications for a variety of ways that resistance is conceptualized and supported as a way of seeking justice in the aftermath of doxxing.

Kaleidoscopic strategies of resistance

As Tamara, Jessie, and Hara's interviews demonstrate, withdrawal from social, digital, and relational spaces is one strategy of resisting against what is expected of victim–survivors in the aftermath of harm. By opting out, pushing back, and refusing to perform contrition or shame, these women created alternative timelines across which the harm of their non-consensual disclosures played out, and where they ended. Withdrawal also took the form of 'scrappy' methods of counter-use of technological tools, such as using platform affordances to perform silence by leaving harassing messages 'on read' as a strategy of defiance and condemnation (Ybarra-Frausto, 1989: 70).

Jessie and Hara's accounts illustrated how deploying silence, or retreating to social relations of care in which silence could be respected, can shore up the autonomy of victim–survivors to participate in the contexts and to the people of their choice. Conceptualizing withdrawal and confrontation not as binary actions of passive or active responses, but as strategies specific to the needs of victim–survivors shows the variety of ways that resistance can be performed. These insights highlight the importance of respecting victim–survivor's choices to perform resistance on their own terms.

Centring victim–survivor voices

The importance of voice was central to this chapter's exploration of complaint and confrontation as other strategies of resistance and autonomy over information. As Sabiyah and Joe's interviews illustrate, confronting the perpetrator of doxxing can lead to healing, resolution, and, in some cases, forgiveness. While it is important to note that victim–survivors of all abuse do not owe their perpetrators forgiveness, the validation of harm which can be possible through confrontation, and the genuine offering of apology is one way in which victim–survivors of doxxing can pursue remedies on their own terms. As Joe's interview illustrated, these remedies may not be accessible to all people, as his perpetrator's apology played out within the context of masculine friendship and camaraderie. This connects back to the findings in Chapter 5, where masculine appeals to honour and validation can result in the mutual respect between men. These complexities and, indeed, different approaches to seeking and receiving justice outcomes affirm the importance of finding a balance between 'valorizing' victim–survivor strength and resilience in the aftermath of harm and holding space for non-linear, faltering forms of healing. As McGlynn and Westmarland (2019: 186) state, this is key to engaging with 'justice [as] a lived, ongoing and ever-evolving experience and process, rather than an ending or result'.

Seeking justice and affirming informational self-determination

The final discussion of resistance in this chapter considered the creative, liberatory, and expansive opportunities to resist against future disclosures of personal, identifying, and sensitive information. As Chelsea's reflections on her experience of being doxxed by a large telecommunications company indicate, personal data is increasingly stored and sold in negligent and insecure ways. The challenges of protecting and advocating for one's personal, identifying, and sensitive information in this context cannot be understated. Affirming the right of all digital subjects to their privacy, to the control and choice over how their information and data is accessed, in what contexts, and by whom is therefore a key opportunity for resisting the harms of doxxing in the future. While some participants desired a criminal justice response after being doxxed, others noted that being heard by their communities (Eden), regaining control over their personal information (Chelsea), or using technologies to craft their own ways of rejecting the harms of doxxing (Hara, Jessie, Tamara) also represent forms of self-defined justice. Informal justice seeking scholars have critiqued the limitations of criminal justice system responses to harm. The implications of how victim–survivors choose to resist after doxxing links up with these critiques and supports further victim-centring research and advocacy for justice on the terms of victim–survivors, rather than as defined by law or the criminal justice system.

There is expansive insight to be gained in exploring the justice needs and responses of victim–survivors operating along a 'continuum of agency' (Dunn and Powell-Williams, 2007: 978). The findings from the interviews of this study reflect strategies of agency along varied – and at times, opposing – axes, such as withdrawal and confrontation, anger and forgiveness, pragmatism, and imagining a future with more expansive justice responses. Engaging with the myriad forms of agency evident in the accounts of victim–survivors does not frame some responses as legitimate, and others as limited, even in moments where the agency of resistance might be contradictory. Instead, these varied examples of agency in the aftermath of doxxing illustrate that the 'kaleidoscopic' pathways to justice also spark a kaleidoscope of possible ways to resist against harm. This gives scope to imagine forms of resistance that are 'constantly refracted through new experiences or understandings' obtained by the victim–survivor (McGlynn and Westmarland, 2019: 179). The 'kaleidoscope' of ways that victim–survivors might embody resistance, therefore, is dynamic, multiple, and subject to the changing needs of those who endure doxxing over time.

7

Conclusion: Affirming the Right to Informational Autonomy

This book has mapped out how doxxing is experienced as a harm that challenges the relational and informational worlds of victim–survivors. By centring the perspectives and needs of victim–survivors of doxxing, we gain insight into the limitations of legal conceptualizations of privacy as negative and restrictive, which people either can or cannot maintain. While doxxing is criminalized in many jurisdictions, victim–survivors of doxxing often lack awareness of anti-doxxing laws and are not supported by law enforcement and police in making legal complaints. Even when criminalized, legal framings of doxxing do not capture the complex harms of doxxing, nor the intersecting justice needs of victim–survivors in the aftermath of non-consensual disclosure.

Instead, this book has unearthed the messy, non-linear, and relational ways in which private lives, personal information, and sensitive aspects of our identities come to represent *our selves*. By bringing the philosophy of information into conversation with criminological and posthuman accountings of harm, I have shown how the harm of doxxing plays out not only through information but within our informational selves, harming through impositions and restrictions on the target's informational autonomy. Doxxing leverages gendered information about bodies, sexualities, medical histories, and more to exploit information as a form of gendered violence and abuse. The ongoing impact of these disclosures causes sexual shaming, reputational harms, ruptures to personal, intimate, and family lives, marring the subject's free enjoyment of consensual disclosure in the future. Just as our histories, bodies, sexualities, and gender identities cannot be divorced from the self, our information cannot be disentangled from our subjectivity. Personal data is enmeshed within the relational, sensorial, and embodied ways that we connect with each other and the world around us. Harms to informational autonomy, therefore, are harms to our social and relational ways of *being*, and cannot be limited to understanding doxxing as an issue of free speech or an example of hate speech.

By developing a conceptual model of the linkages between doxxing, privacy abuse, and their impacts upon informational autonomy, this book affirms the importance of the relationship between privacy and information. By affirming that privacy is key to the full enjoyment of informational autonomy, doxxing can be understood as a harm that impacts the possibilities of self that we are in the process of 'becoming' (Braidotti, 2019c: 36). Framing doxxing within these contexts extends privacy abuse beyond a static, linear event to one that fractures into ongoing and enduring risks of future disclosure. Doxxing can also spark affirming and self-defined strategies of resistance for victim–survivors who develop new ways of reclaiming autonomy over their information after its non-consensual disclosure. These strategies of resistance have also contributed to understandings of how victim–survivors desire and seek out justice in the aftermath of experiencing harm and abuse.

To bring these reflections to a close, this chapter provides a brief overview of the key contributions of each chapter. The implications of the findings of this study are considered in detail, with future directions for research, policy, and professional practice earmarked. I conclude by showing how doxxing as a form of privacy abuse constitutes an ongoing, enduring, and relational harm. Framing doxxing in this way requires that the rights to autonomy over personal, private, and sensitive information are taken seriously. I conclude by calling for important conceptual and theoretical shifts toward an affirmative model of informational autonomy.

Research implications

Doxxing takes on many motivations, impacts, and forms which are endured by victim–survivors. The harms of doxxing extend far beyond the initial moment of non-consensual disclosure into ongoing, cyclical, and imbricated harms which place victim–survivors at risk of future disclosure of their information. In Chapter 2, I argued that doxxing is not an umbrella term to describe other digital harms but a unique harm in its own right which impedes the free, autonomous participation of subjects in both digital and offline life. Further, doxxing not only coincides with gendered identities and power dynamics but is a performance *of* those identities and dynamics in digital spaces, which draws attention to the unique affordances of platforms which sustain and permit harassment and abuse of personal information. These framings led to my critique of the *relational rupturing* of doxxing in Chapter 3, where personal information abuse harms victim–survivors not only in their present social communities but in the potential social relations they may form in the future. For the women and men interviewed in this study, doxxing was used to shame, degrade, and intern them within the patriarchal hierarchy of worth. These findings show the complexity of how patriarchy not only harms women but all people in their relation to each other.

I explored the first-and-second order impacts of doxxing further in Chapter 4, mapping how doxxed information continues the life cycle of data abuse for women and children, who experience an additional dimension of shame, risk, and vulnerability. In Chapter 5, I considered the unique gendered experience of harm for the men of the study, who reported personal information abuse as an affront to their masculinity and status within peer networks. While the men in this study had different strategies of support and help-seeking after experiencing harm, they too were impacted by patriarchal judgements of worth and requirements to be 'strong' after experiencing data abuse. These findings led to a consideration of what *informational autonomy* might offer victim–survivors as a strategy to remake and resist intrusions to personal identity. By affirming the right of all people, but especially victim–survivors of doxxing, to enjoy full control over how, when, and to whom their information becomes known, strategies of remaking and reclaiming personal data in the aftermath of doxxing are possible. As such, the findings and theoretical developments made in this study present important implications for how we understand, research, and address issues of gendered, interpersonal, intimate, and family violence, as well as technology-facilitated violence and digital harms.

Highlighting the role that abuse of personal, identifying, and sensitive information plays in tactics of control, harming, and manipulation presents several important implications for the ways we research and craft policy responses to interpersonal violence. Many instances of doxxing can be identified as a means of perpetrating family, domestic, and intimate partner violence. These findings are particularly important when considering that the 'life cycle' of personal data outlives all relationships (Ambrose, 2012). Personal data remains vulnerable to abuse regardless of the status of the relationship. This might sharpen our awareness of the sensitivity of personal information in all kinds of relational contexts – for example, how might we support victim–survivors who are co-parenting with an abuser, who has access to their/children's personal/medical/location-based information? Including doxxing in identifying domestic and family abuse is an important future step to sharpen efforts to safe-guard data regardless of the status of the relationship.

The information exploited to dox most of the women in this study was operationalized in ways that shamed them, denigrated their sexual lives, and positioned them as 'less than' within the hierarchy of worth which underscores patriarchal value judgements. Importantly, these experiences of shame were often not independent from domestic and family violence but intersected with it. Many of the women reported that their doxxing was perpetrated by an ex-partner to embarrass and humiliate them in their broader social, family, and professional networks. This illustrates how the doxxing of personal information compounds across gendered lines, impacting the victim–survivor's ability to freely participate in social contexts of their choosing. This is the imbricated, cyclical, and compounding nature of doxxing, which cannot be classified as one discrete instance of family, domestic, interpersonal, or technology-facilitated abuse. The harms of non-consensual disclosure refract across many aspects of the victim–survivors' lives and identities. This provides new insights into the ways that gendered, sensitive information is leveraged to coerce, manipulate, and abuse victim–survivors in patterns of control.

A key contribution arising from this study is that gender impacts and shapes the forms, harms, and experiences of doxxing. By framing doxxing as a form of digital gender-based violence, I contribute to growing critical scholarship on the ways technologies invite and afford abuse which manipulates, coerces, and harms victim–survivors (Wood et al, 2023; Wooley, Iliadis and McMahon, 2023; Brown and Hegarty, 2024). These framings also expand prior research which connects privacy abuse and surveillance to tactics of digital coercive control. Digital coercive control researchers (see Huber and Godfrey, 2024) have identified the use of smart devices, social media, and other digitally mediated technologies as tools which domestic and family violence perpetrators use to intimidate, harass, surveil, and manipulate their victims. It is not surprising that for many of the women interviewed in this study, their known or suspected perpetrator was an ex-partner who intended to shame and harm them after their relationship ended. Lever and Eckstein (2020: 60) note that victim–survivors of digital coercive control report that their experiences of abuse post-relationship continue in 'new, amplified ways brought about by post-dissolution norms'. The experiences of Carol, Hara, Odette, Sabiyah, and Jessie show how doxxing amplifies the harms of digital coercion by exploiting the virality, visibility, and scale of social media where their information was shared. The findings from these interviews offer important insights into how doxxing and digital coercive control coincide with broader abusive dynamics where a perpetrator intends to intimidate and degrade their target using personal information. As such, these findings expand understandings of how a range of abusers, including domestic, family, and intimate partner abusers, use doxxing as a means to control, manipulate, and harm their victims.

In addition to family, domestic, and intimate partner violence, doxxing is also deployed as a form of interpersonal violence. As DeKeseredy and Schwartz (2005: 355) note, there are many issues in recognizing interpersonal violence, as 'there are many behaviours that may seem extremely violent but nevertheless are not viewed that way by many or most people'. The accounts of Tamara, Tegan, Joe, and Andrew provide important expansions on what is known about doxxing perpetrators and their motivations, as they were doxxed by friends, classmates, and colleagues rather than intimate partners or family members. For Tamara and Tegan, they were doxxed by close female friends who showed a 'complicity in their own subordination' by patriarchy in their efforts to sexually shame and condemn them (Knowles, 2021: 1318). For Joe and Andrew, being doxxed by friends and classmates demonstrates how young men leverage the 'uncertainty of rank and porous dividing lines between different types of masculinities' to assert power, strength, and dominance over each other (Ozyegin, 2018: 244). Another example is the doxxing of Molly's child, who was targeted alongside Molly in an anonymous attack. While the anonymous perpetrator may have intended for the doxxing of her child to intimidate and distress Molly, the dynamics of disclosure in this case do not neatly fit into a perpetrator–victim model that can be characterized as family, domestic, or intimate partner violence. Expanding understandings of doxxing also generates new understandings of the types and complexities of interpersonal violence, which include sensorial and relational harms to identity (see Donovan, Magić, and West, 2024) reputation, status, and security within relationships. Understanding doxxing in platonic and non-familial contexts expands understandings of interpersonal violence beyond perpetrator–victim dynamics in family, domestic, and intimate partner violence.

As technological interventions in daily life continue to develop, the importance of understanding the role of technology in mediating doxxing and privacy abuse grows ever more important. Carol's account illustrates how 'unanticipated harmful effects' might arise, as she was doxxed when confidential work documents assigned to her care were leaked to her boss anonymously (Wood et al, 2023: 1387). These documents, ordinarily encrypted through a cloud-based secure site, should not have been accessible to anyone but Carol. Thus, the breach of this encrypted system not only facilitated the doxxing of confidential information under Carol's control but extended the use of technology to harm by damaging Carol's reputation and credibility at work. The intersection of doxxing here as both a mediated act through technological capabilities and a harm experience afforded through an unintended aspect of its design presents a new way to think about the role of technology in moments of harm. Insights from this study therefore contribute to growing scholarship distinguishing between the *facilitation* of harm and abuse through technology and the *affordances* and *invitations* offered by technology. Further research into doxxing – and indeed, other forms of gender-based and digital abuse – can contribute to these theoretical debates and innovations in the future.

I have also developed a new theoretical framework of informational autonomy to understand *how* the harm of doxxing is experienced. Doxxing harms are experienced as harms to security, to the body, to the self, and extend beyond a linear framing of time. A crucial impact of doxxing is the risk of harm it sustains for the individual's unimpeded control over their information in the future. As such, I echo Eckert and Metzger-Riftkin

(2020), Corbridge (2018), Igo (2018), and Nissenbaum's (2004, 2018) call for contextual integrity. Stephenson et al (2025: 1) have recently outlined that 'digital autonomy' is linked to victim–survivor's ability to exercise their rights to 'economic mobility [and] feelings of safety' in digital spaces. Drawing on these theoretical framings of autonomy, and the nuances of navigating the diverse contexts of digital life, I have developed the concept of informational autonomy – an extension of the importance of contextual integrity and self-determination – to provoke an affirmative model of individual privacy rights.

Informational autonomy scopes out privacy as the positive enjoyment of liberty, freedoms, and self-expression, rather than the negative and restrictive model of a right to be possessed, and therefore lost. While Ethan's account of 'self-doxxing' was omitted from the discussion of this text, his reflections on the affirming, community-building potential of sharing personal information in crypto-NFT forums underscores what is *possible* when informational autonomy is retained (see further account of Ethan's experience of self-doxxing in the *Glossary of victim–survivor experiences*). An affirmative model of informational autonomy can reinvigorate research both within and beyond digital harms and abuse scholarship, and is aligned to support policy and law interventions. For example, there are growing concerns about the digital rights of children and their safety in online and terrestrial spaces across the lifespan. In many jurisdictions, including the United Kingdom, Australia, and the EU, the digital rights of children have been signposted as key priorities for governmental intervention and review (see Ofcom, 2024; OECD, 2024). I contribute to this growing work by contextualizing doxxing as a form of privacy abuse which sustains tactics of control over children, who may be targeted in both explicit and tacit efforts to harm adult victim–survivors. The concept of informational autonomy is a unique contribution to theorizations of privacy, digital rights, and selfhood, and has scope to invigorate user experiences on digital platforms, as well as in policies across both government and private sectors.

Not only do I develop a critique of the limited ways in which individual rights to privacy are articulated by law, I also critique the limitations of seeking justice within the criminal justice system. Legal rights to privacy, where they exist, exalt the 'right to be let alone' or to enjoy private spaces, typically within the home. Even where more nuanced laws, such as the European Union *General Data Protection Regulation*, set out the rights to informational self-determination, these rights are not unitary and capture limited ways that they may be enjoyed. As such, much legislation fails to conceptualize the rights to privacy that might be enjoyed across digital and 'onlife' experiences. In February 2024, the Australian Prime Minister, Anthony Albanese, announced that 'anti-doxxing' legislation would be drafted as part of the ongoing process of amending the *Privacy Act (1988)*.

While amendments to the *Privacy Act* are indeed a step forward in supporting victims of doxxing in Australia, a great deal of advocacy is needed to ensure that the rights of individual users are centred in legislation. Historically, privacy amendments in Australia have granted law enforcement and government agencies further power to surveil and control the population, particularly through access to metadata (Wilson, 2020). As Zuboff (2023) cautions, user-rights to data privacy, informational autonomy, and protection from digital surveillance are antithetical to late-stage capitalism which has modelled an extractive model of digital life where users are the product, not protected. Ineffective (or non-existent) legal framings of privacy rights and abuse therefore do not leave victim–survivors of doxxing with sufficient avenues to seek redress, justice, and the securitization of their sensitive and personal information.

Jannice Käll (2024: 201) has also cautioned that legal frameworks of 'rights have been criticized as being too conceptual and abstracted from the actual expression of power that orders bodies', calling instead for critical interrogations of the power within law. This is a key implication of the findings arising in this book: victim–survivors of doxxing need robust, well-resourced, and empathetic support services and law enforcement responses should they choose to pursue justice through the criminal justice system. I do not tout law reform of privacy rights as the silver-bullet for reckoning with doxxing, nor do I silo justice seeking as only legitimate when accessed through the traditional justice system. However, it is pivotal that victim–survivors are taken seriously and offered meaningful support when they do choose to pursue a criminal justice complaint. The critiques set out in this study – particularly of ineffective and incomplete investigations and care provided by police and other law enforcement – present implications for further victim-centred, empathetic, and skills-oriented training for police officers and personnel. This begins with a skilling-up of officer understandings of individual privacy rights and protections against doxxing. Most importantly, police must engage with complainants of any form of abuse, doxxing or otherwise, with care and respect.

Victims of doxxing deserve to be taken seriously by police and for meaningful progression and investigation into their complaint to occur. This would likely entail training into digital evidence gathering and tracing of personal information shared without consent, a strategy of evidence collection which also requires police agencies to have adequate tools and equipment (see Cross et al, 2021). While police training and up-skilling into proper and rigorous investigations into privacy abuse is needed, it is also critical to consider that some members of law enforcement agencies themselves perpetrate doxxing and privacy abuse upon members of their communities. The *Commission of Inquiry into Queensland Police Service responses to domestic and family violence* (2022: 176, 178, 313) found that some officers who perpetrate domestic, family, and intimate partner violence inappropriately and illegally access police satellite surveillance and databases containing the personal information, home, and workplace addresses of their victims. Similar findings in other states and territories of improper and illegal police misconduct involving the personal information of civilians have been reported.[1] Other high-profile cases of police doxxing include the outing of Dani Laidley,[2] a trans woman who was doxxed when officers leaked photographs of her in a police station before she had come out publicly, and Officer Glen Punchard's hacking of a police database to share a domestic violence victim–survivor's new home address with her abuser.[3] As such, addressing justice responses to doxxing also requires police officers who dox their victims to be held to account (see Anderson, Farmer, and Tyson, 2024).

The engagement with victim–survivors experiences of doxxing throughout this book has followed in the qualitative, feminist tradition of foregrounding what is relational, sensorial, and deeply personal to experiences of harm and abuse (Panfil and Miller, 2015; Westmarland and Bows, 2018). By embracing the deep textures of conversations through narrative and de-hierarchized interview methods, the lived experiences of the 18 women and men have informed the theoretical insights of this book with rich and nuanced detail. However, it is important to note that by drawing on a small sample of victim–survivors of doxxing, this book does not capture personal data disclosures that play out in different contexts and relations to technology. The cohort of participants, while diverse in location, socio-economic status, race and cultural backgrounds, does not represent a complete accounting of victim experiences.

For example, a notable (and unintended) exclusion from the participant pool of this study is the perspectives of trans and gender-diverse people, who experience the doxxing of their personal, identifying, and sensitive information in social and political climates that are increasingly volatile and targeting (Anderson, 2025). There is also significant scope in future research to consider the life cycle of personal information, and the stages of childhood, adolescence, and adulthood where victim–survivors may be particularly vulnerable to the abuse of their data through doxxing. As with many qualitative research projects, the benefits of deep diving into personal, subjective, and nuanced participant experiences also requires balancing against over-generalization or omission of different experiences than those captured in a small cohort. As Kalu and Bwalya (2017: 50) suggest, the benefits of qualitative studies come from the richness, depth, transferability, and contextual relevance of the data, rather than 'some grand statistical claim' to significance. While this book is not a representative, quantitative snapshot of doxxing experiences globally, it sets out a framework for future research agendas: a framework of data abuse which is critical of and attentive to the need for liberatory, affirming and radical steps toward informational autonomy. Some of these steps are outlined in the following manifesto, which calls for scholarly, policy, and activist efforts to protect the informational autonomy of all people, in all social, offline, and digital spaces.

Towards an affirmative ethics of informational autonomy

Throughout this book, I have demonstrated the need for an alternative to the inhibiting, restrictive models of legal privacy rights which are not presently guaranteed or articulated consistently. The victim–survivors of doxxing interviewed in this study were not taken seriously by law enforcement and did not access law as a genuine site of justice. Clearly, an alternative approach to advocating for privacy rights is needed, which affirms the care, respect, and custodianship that personal, identifying, and sensitive information deserves. Later in this chapter, I consider how affirming the autonomy to control and choose how information is used, by whom, and in what contexts can be framed as an expansive and liberatory practice.

In Chapter 2, I discussed Braidotti's (2019b: 470) affirmative relations and ethics as an emancipatory process of affirming 'our innermost freedom – the freedom to become all we are capable of, all our bodies can take'. In contrast to the limitations of legal privacy rights which understands harm as something which takes from the subject, Braidotti's affirmative ethics maps out what our freedom from harm can and should be imagined to be. Affirmative ethics, and its resulting political action, 'draws on a positive notion of power through which reality could be enacted differently' (Postma, 2016: 312). Braidotti's framing of ethics and power, therefore, considers what might be liberatory about coming into ethical informational care through *relationality* to all actors. This also aligns with Floridi's (2013: 132) view that information ethics requires 'a mature sense of ethical responsibility for and stewardship of [the] whole environment, both natural and artificial (that is, in one word, informational), to foster responsible care of it rather than despoliation or mere exploitation'. This connects to the relationality we share within the informational environment as it situates 'stewardship' as something extended to the *whole* environment, rather as an individual responsibility. Fostering ethical values toward information, privacy, and our contextual integrity and autonomy shows respect for the information of our selves – an ethical value framework that stands in opposition to the extractive and depersonalized data models of digital life today.

Importantly, calling for the right to informational autonomy to be articulated, protected, enhanced is not a call for 'yet another law/mandate' which falls to the wayside. Käll's (2024: 202) critique of anthropocentric human rights similarly identifies that bodies that are recognized through law are 'equipped not only with conceptual rights but with the power to fulfill their capacities'. In thinking about protecting informational autonomy, this affirms the generative potential/power that informational selves already possess. This also connects back to Braidotti's encouragement to affirm 'all that we can be, all that we can take' in the pursuit of informational freedom and liberation. This leads to my rejoinder that affirming informational autonomy is not only an antidote to data harms and abuse but an expansion of the liberatory modes of relation, expression, connection that are possible when we are fulling affirmed in our informational autonomy. Autonomy over information is power, is connection, is self-writing, is self-producing. It is the power to craft and express all that we can be, all that we are capable of, all that our bodies can take.

Creativity and liberation in the aftermath of doxxing

How might informational self-determination be liberatory? Informational self-determination provides another option for victim–survivors, moving beyond the ideal that they can 'take back' control over their doxxed information. Instead, the liberatory potential of informational self-determination sparks a new awareness of understanding the self *as* information – a new way of situating selfhood within information not only online, but in all aspects of life. As I discussed in Chapter 2, there are a raft of affirming ways that we can perform our subjectivity through digital mediums, meeting 'the need for self-expression, the performance of identity [and] the chance to reinvent oneself' in new ways online (Floridi et al, 2015: 11). Self-expression, reinvention, and the ways that we perform our identities are entangled with the ways we access and utilize our information in both offline and digital contexts. What is liberatory about informational autonomy is its capacity to help subjects understand their personal information in the 'onlife' and advocate for their informational rights and boundaries in these contexts.

As such, we might consider that the harm outcomes of doxxing can be 'creative and affirming' alongside their negative impacts (Green, Calverley, and O'Leary, 2021: 853). Doxxing spotlights the aspects of personal information that might once have been taken for granted. The harm of non-consensual disclosure 'creates' a different appreciation of its worth by victim–survivors, and puts forward a new baseline for working through what informational self-determination might look like. While living with the violence of doxxing might feel unrelenting, where the risk of ancillary forms of harms looms large, a liberatory way of understanding resistance post-doxxing is to think about what is 'opened up'. This 'opening up' can be read through what Didi-Huberman (cited and translated by Rouvroy, 2013: 148) suggests as 'taking position' upon the histories of the self in the past into the present; 'one must also take into account everything one leaves aside, the off-frame that exists behind us … situating oneself in the present and aiming at the future'. For victim–survivors of doxxing, the 'off-frame' of non-consensual disclosure is brought into the present and can extend into a future where they 'take position' over their personal information.

Reclaiming privacy can affirm what is shared, knowable, and visible after harm. Regardless of the form this takes, this process must be on the terms of the victim–survivor. Schneider (1991: 996) terms this the 'affirmative potential of privacy', where the 'autonomy, decisional privacy … and affirmative self-expression' of the subject is respected. While the harms arising from doxxing are indisputable, Tegan's account also models how post-traumatic growth provides a parallel way of understanding how victim–survivors experience the aftermath of harm. In contrast to post-traumatic stress or other responses to harm, post-traumatic growth considers the '"meaningful transformations" that can be obtained from trauma and other adverse events' (Jayawickreme et al, 2021: 148). Positive growth might be conceptualized in a variety of ways in the aftermath of harms like doxxing. For participants like Eden, the ability to share her experiences of doxxing with other people in her community and social groups generated close connections and support which she found meaningful. For participants like Chelsea, growth in the aftermath of doxxing took the form of a pragmatic review of the contexts in which her personal data was stored. For Tegan, she found a new way to self-define the value of her information as something she kept secret and 'close' to her.

In line with Green, Calvary, and O'Leary's (2021) framings of growth and resilience not being used to conceptualize victim–survivor responses as 'good' or 'bad', I take up the post-traumatic growth framings here to note how harm and growth can coincide, often in the same moment. Tedeschi, Calhoun, and Groleau (2014: 505) note a similar paradox, where 'the maintenance of the growth experienced may require unpleasant periodic cognitive reminders of what has been lost, so that in an apparently paradoxical way, what has been gained remains in focus'. For Tegan and other participants who have grown and changed in the aftermath of being doxxed, the harm of the non-consensual disclosure of their information sits parallel to what they have deemed meaningful in its aftermath. This underscores what I discussed in Chapter 6 as the multiplicity of harmful experiences sitting alongside loss, change, and growth.

It is also important to note that renewed forms of informational self-determination can be interpreted as liberatory as opposed to merely reacting to fear or the perceived risk of future doxxing. Floridi (2011: 562) suggests that constructing the self through technology presents an opportunity to be 'thriftier' and 'more skillful' in our sensibilities about how we participate and exist in digital spaces. In Chelsea's case, her renewed commitment to and interest in where her data is stored, how it is used, and by whom is one such way of 'tinkering with the self' toward the liberation of her data from the control of other parties (Floridi, 2011: 565). Tegan's choice to 'keep a lot of secrets' in the aftermath of doxxing also highlights a change in her practices of how she discloses information about herself in the aftermath of harm. As Schneider (1991: 999) critiques, however, informational self-determination must also be backed up by changes to the ways we recognize privacy. She notes that 'privacy that is grounded on equality, and is viewed as an aspect of autonomy, that protects bodily integrity and makes abuse impermissible, is based on a genuine recognition of the importance of personhood' (Schneider, 1991: 999). Victim–survivor interpretations of informational self-determination are likely to vary, but are unified by a shared desire to possess and exercise autonomy in ways that affirm their personhood, rather than respond to harm.

There is an opportunity to turn the framework of data liberation on its head to serve the interests of victim–survivors of doxxing. Presently, data liberation is a practice employed by businesses, health organizations and governments to mobilize cached data to generate new insights and competitive market advantages (Westergren, 2016; Afful-Dadzie and Afful-Dadzie, 2017; Crooks, 2022). Framings like 'making the data work for you' situate this practice squarely in what Zuboff (2022: 3) identifies as the 'commodification of human behaviour' through surveillance, where all data can be produced toward profit. But this term can also be co-opted to situate individual data insights. What might it mean for victim–survivors to reflect back on their 'cached' experiences online? To consider the various ways that their personal data and information has been made accessible, both explicitly and with their permission, and implicitly without their awareness? What do their reflections on the inefficiency of reporting data abuse to social media and other digital platforms lead to? These 'cached' experiences have been 'mobilized' by the participants of this study in their reflections, leading them to new insights, commitments to security, language, and vocabulary about how they wish to appear online, to whom, and in what contexts. That is to say, that victim–survivors of doxxing can 'liberate' their past voluntary disclosures of information to now attend to their *new* objectives and needs.

Data liberation, increased transparency over data storage and security protocols, enhanced protections for users and their data on social media platforms; there are many forms through which the informational autonomy of digital subjects can be affirmed. In the following section, I develop a manifesto of sorts, in the styling of Uldbjerg's (2021b) manifesto on 'writing victimhood' as an intrinsic aspect of conducting activist and participatory research. Framing a manifesto of the forms an affirmative ethics of informational autonomy *could* take, therefore, serves as a commitment to the ongoing work, activism, research, and learning in this space. Bringing Braidotti's (2019b: 470) call for 'the freedom to become all we are capable of' into a manifesto to imagine what freedom from privacy abuse looks like is therefore a political act. It requires an imagination of models of care, respect, and ethical treatment of information of *all* entities. Freedom from privacy abuse stands in opposition to the systems of capitalism and patriarchy which underscore how information is valued, leveraged, and abused. Framings and conceptualizations of these values do exist: through data sovereignty and its condemnation of colonial and imperial harms, control, and management of Indigenous peoples (Walter et al, 2021; Walter and Suina, 2023); through open data philosophies which call for the shattering of obscured, shadowing data storage (Tegmark, 2018); through data provenance, which suggests strategies of shedding light on data's origins, its true owners and how it extends into broader use (Fordyce and Jethani, 2021). Understanding how doxxing harms is key to the development of this framework as it considers not only how privacy abuse is a conduit for other forms of harassment but how a revitalized ethics of respect and care for information might open up liberatory and expressive modes of relation in the digital futures to come. With inspiration from the values and political commitments of these framings, I develop the following manifesto.

Affirming the right to informational autonomy

The right to contextual autonomy: Subjects deserve the choice of how, when, why, and to whom their information is disclosed. This relates to a respect and value of information as intrinsic to each and every subject. No one entity's information is more or less worthy of these protections than another.

The right to opt in and out of public domains: Desires for presence and absence in social spaces, both terrestrial and digital, shift throughout the lifespan. There is a need for greater protections for the data that is presently 'left behind' – for example, finding ways to scrub and encrypt personal and identifying information when terminating a service with a telecommunications company, a hospital, or a company.

The right to respect and care for all information: We must pinpoint the platforms, sites, and companies that presently profit from the sharing of sensitive information and content. Platforms should be liable for the non-consensual sharing of information on their sites. This has implications for a wide range of abuses including and beyond doxxing.

The right to privacy across the lifespan, beginning in infancy and extending to end of life: Inspired by Uldbjerg's (2023: 536) critique that digital sexual assault is considered a 'young person's issue' which silos the experiences of adults at all stages of life as unintelligible, privacy rights must be articulated as rights for all. This has implications for those who cannot speak for themselves or actively make choices about how their information is captured and shared.

This manifesto toward an affirmative model of informational autonomy is a starting point, rather than an arrival. Its aims, values, and objectives are ambitious – rather than this being a point of critique or an issue of feasibility, its ambition serves to push back against the accepted models of data extraction, lack of care and the sustaining of platforms which exploit and permit abuse. Critiquing the capitalist, patriarchal, and extractive models which have presently permitted these harms is arduous and difficult work – but it is work worth doing. As our social lives continue to intersect with and respond to the technologies which develop around us, we too can shape and inform how the rights of all informational entities are secured, respected, and affirmed in these spaces.

Conclusion

In this book, I have shown that doxxing exploits personal, identifying, and sensitive information in enduring, compounding, and complex ways. The findings outlined in this book indicate that doxxing is a poorly understood harm which is often conflated with group politics and the performance of identity in online spaces. Significantly, the findings also illustrate that victim–survivors experience the harm of doxxing as a rupturing of their social lives which limits their ability to freely participate in the contexts of their choice. The ruptures which doxxing produce grant insight into how subjects are constituted, embodied, and performed through information. When ruptures arise, the ability to choose how, when, and to whom the self can be known is denied. These ruptures demand attention be paid to the ontology of information and its connection to digital subjectivity. Importantly, the findings of this research also indicate that doxxing is a form of digital gender-based violence, which results in complex ways that non-consensual disclosure is leveraged to place women, men, and children within the patriarchal hierarchy of worth.

The findings of this book are significant because they shift understandings of privacy and anonymity from legal or cultural thresholds to something that is entangled with how and who we are. Adopting Braidotti's framing of the affirmative potential of 'freedom,' we might understand our information as key to the ways that we conceptualize, protect, and celebrate privacy as an ontology of digital subjecthood. Our social lives, ways of knowing, modes of connection, and communication are intertwined with technology, more so now than ever. The issue of privacy abuse and the perpetration of doxxing as a form of this abuse require rigorous and ongoing research, critique, and care. By affirming privacy as an ontological aspect of the digital self, deserving of respect, care, and protection, we come closer to finding balance with the ways our information structures and shapes our digital selves, lives, and worlds around us.

Notes

Chapter 1

1 The names of the people who were interviewed throughout this book have been pseudonymized to protect their anonymity.

2 Also referred to as 'doxing'.

3 See *Wetboek van Strafrecht*, 1997 (Netherlands).

4 See *Privacy and Other Legislation Amendment Bill*, 2024 (Commonwealth of Australia).

5 See *Information and Electronic Transaction (IN) Law*, 2016 (Indonesia).

6 See *The Personal Data (Privacy) Amendment Ordinance*, 2021 (Hong Kong).

7 See *Harmful Digital Communications Act*, 2015 (New Zealand).

8 See *Cybercrimes Act No. 19 of 2020* (South Africa).

9 See *Protection from Harassment Act*, 1997 (United Kingdom).

10 See *Malicious Communications Act*, 1998 (United Kingdom).

11 Previously known as Twitter.

12 Human Research Ethics Committee Approval ID: 2021-20401-15573-3.

Chapter 5

1 The account of Ethan has been excluded from analysis, as he self-defined doxxing as an affirming practice to build trust within cryptocurrency and NFT communities (see *Glossary of Victim–Survivor Experiences*).

Chapter 7

1 Bogel, A. and Knaus, C. (2024) 'Revealed: The amount of times Australian police have breached the trust afforded them', *The Guardian*, 27 June. Available from: https://www.theguardian.com/australia-news/article/2024/jun/28/revealed-the-amount-of-times-australian-police-have-breached-the-trust-afforded-to-them#:~:text=A%20Guardian%20Australia%20investigation%20has,internal%20police%20databases%2C%20since%202020.

2 Geary, M. (2021) 'IBAC investigates after police allegedly share photo of former AFL coach Dani Laidley with "vile" comments', *The Guardian*, 10 November. Available from: https://www.abc.net.au/news/2021-11-10/danielle-laidley-ibac-appeal-police-photo-allegation/100610762.

3 Smee, B. (2021) 'Queensland police officer who leaked domestic abuse victim's address resigns', *The Guardian*, 4 October, https://www.theguardian.com/australia-news/2021/oct/04/queensland-police-officer-neil-punchard-who-leaked-domestic-abuse-victims-address-resigns.

References

Ackelsberg, M.A. and Shanley, M.L. (2018) 'Privacy, publicity, and power: A feminist rethinking of the public-private distinction', in N.J. Hirschmann and C. di Stefano (eds) *Revisioning the Political*, Routledge, pp 213–33.

Adair, J. and Senn, C.Y. (2025) 'The role of rape myth acceptance, situational context, and gender in individual's perceptions of image-based sexual abuse victims and perpetrators', *Journal of Interpersonal Violence*, 08862605251319022.

Afful-Dadzie, E. and Afful-Dadzie, A. (2017) 'Liberation of public data: Exploring central themes in open government data and freedom of information research', *International Journal of Information Management*, 37(6): 664–72.

Ahmed, S. (2010) 'Foreword', in R. Ryan-Flood and R. Gill (eds) *Secrecy and Silence in the Research Process: Feminist Reflections*, Routledge, pp xvi–xxi.

Ahmed, S. (2014) *The Cultural Politics of Emotion*, Edinburgh University Press.

Ahmed, S. (2016) 'An affinity of hammers', *Transgender Studies Quarterly*, 3(1–2): 22–34.

Ahmed, S. (2021) *Complaint!* Duke University Press.

Albury, K. (2017) 'Just because it's public doesn't mean it's any of your business: Adults' and children's sexual rights in digitally mediated spaces', *New Media & Society*, 19(5): 713–25.

Ambrose, M.L. (2012) 'It's about time: Privacy, information life cycles, and the right to be forgotten', *Stanford Technology Law Review*, 16: 101–54.

Anderson, B. (2025) 'Doxxing to destroy: The convergence of transphobic hate speech and non-consensual disclosure on X', *Crime, Media, Culture*, https://doi.org/10.1177/17416590251345736.

Anderson, B. and Wood, M.A. (2021) 'Doxxing: A scoping review and typology', in J. Bailey, A. Flynn, and N. Henry (eds) *The Emerald International Handbook of Technology-facilitated Violence and Abuse*, Emerald Publishing, pp 205–26.

Anderson, B. and Wood, M.A. (2022) 'Harm imbrication and virtualised violence: Reconceptualising the harms of doxxing', *International Journal for Crime, Justice and Social Democracy*, 11(1): 196–209.

Anderson, B., Farmer, C., and Tyson, D. (2024) 'Police-perpetrated domestic and family violence: A scoping review of Australian and international scholarship', *International Journal for Crime, Justice and Social Democracy*, https://doi.org/10.5204/ijcjsd.3582.

Anderson, B., Lundberg, K., and O'Donnell, S. (2024) 'Embracing the mess in feminist research: Insights into posthumanism', *Qualitative Inquiry*, https://doi.org/10778004241295743.

Anderson, P.S. (2016) 'When justice and forgiveness come apart: A feminist perspective on restorative justice and intimate violence', *Oxford Journal of Law and Religion*, 5(1): 113–34.

Antonsdóttir, H.F. (2020) 'Injustice disrupted: Experiences of just spaces by victim-survivors of sexual violence', *Social & Legal Studies*, 29(5): 718–44.

Apthorpe, N., Shvartzshnaider, Y., Mathur, A., Reisman, D., and Feamster, N. (2018) 'Discovering smart home internet of things privacy norms using contextual integrity', *Proceedings of the ACM on Interactive, Mobile, Wearable and Ubiquitous Technologies*, 2(2): 1–23.

Arceneaux, P.C. and Dinu, L.F. (2018) 'The social mediated age of information: Twitter and Instagram as tools for information dissemination in higher education', *New Media & Society*, 20(11): 4155–76.

Bailey, J., Henry, N., and Flynn, A. (2021) 'Technology-facilitated violence and abuse: International perspectives and experiences', in J. Bailey, A. Flynn, and N. Henry (eds) *The Emerald International Handbook of Technology-facilitated Violence and Abuse*, Emerald Publishing, pp 1–17.

Barnwell, A. (2019) 'Family secrets and the slow violence of social stigma', *Sociology*, 53(6): 1111–26.

Barrett, T. (2018) 'Bourdieu, hysteresis, and shame: Spinal cord injury and the gendered habitus', *Men and Masculinities*, 21(1): 35–55.

Barry, P.B. (2020) 'Doxing racists', *Journal of Value Inquiry*, 55(3): 457–74.

Bellaby, R.W. (2023) 'Leaks: From whistleblowing to doxxing', in R.W. Bellaby (ed) *The Ethics of Hacking*, Bristol University Press, pp 73–93.

Belknap, J. (2010) 'Rape: Too hard to report and too easy to discredit victims', *Violence Against Women*, 16(12): 1335–44.

Benjamin, J. (2022) 'Recognition in the face of harm: Implicated subjectivity and the need for acknowledgment', in R. Kabasakalian-McKay and D. Mark (eds) *Inhabiting Implication in Racial Oppression and in Relational Psychoanalysis*, Routledge, pp 52–75.

Bhagawati, M. and Priya, K.R. (2025) 'Suffering and radical healing among women survivors of acid attack in India: A qualitative study using an augmented reality comic', *Qualitative Research in Psychology*, 22(2): 1–26.

Bignall, S. and Braidotti, R. (2019) 'Posthuman systems', in R. Braidotti and S. Bignall (eds) *Posthuman Ecologies: Complexity and Process After Deleuze*, Rowman and Littlefield Publishing, pp 1–16.

Bluett-Boyd, N. and Fileborn, B. (2014) *Victim/Survivor-Focused Justice Responses and Reforms to Criminal Court Practice: Implementation, Current Practice and Future Directions* (Research Report No. 27), Australian Institute of Family Studies.

Boehnlein, T., Kretschmar, J., Regoeczi, W. and Smialek, J. (2020) 'Responding to stalking victims: Perceptions, barriers, and directions for future research', *Journal of Family Violence*, 35: 755–68.

Bogel, A. and Knaus, C. (2024) 'Revealed: The amount of times Australian police have breached the trust afforded them', *The Guardian*, 27 June. Available from: https://www.theguardian.com/australia-news/article/2024/jun/28/revealed-the-amount-of-times-australian-police-have-breached-the-trust-afforded-to-them#:~:text=A%20Guardian%20Australia%20investigation%20has,internal%20police%20databases%2C%20since%202020.

Briandana, R., Oktavianingtyas, I., and Marta, R.F. (2020, November) 'Cybercrime in online dating site: Pornography business in the virtual world', in *ICSSED 2020: The Proceedings of the 4th International Conference of Social Science and Education, ICSSED 2020, August 4–5 2020, Yogyakarta, Indonesia*, pp 419–25.

Braidotti, R. (2019a) *Posthuman Knowledge*, Polity Press.

Braidotti, R. (2019b) 'Affirmative ethics and generative life', *Deleuze and Guattari Studies*, 13(4): 463–81.

Braidotti, R. (2019c) 'A theoretical framework for the critical posthumanities', *Theory, Culture & Society*, 36(6): 31–61.

Bennett, J. (2010) *Vibrant Matter: A Political Ecology of Things*, Duke University Press.

Brison, S.J. (2002) *Aftermath: Violence and the Remaking of a Self*, Princeton University Press.

Brison, S.J. (2017) 'Personal identity and relational selves', in A. Garry, S.J. Khader, and A. Stone (eds) *The Routledge Companion to Feminist Philosophy*, Routledge, pp 218–30.

Brown, S. (2006) 'The criminology of hybrids: Rethinking crime and law in technosocial networks', *Theoretical Criminology*, 10(2): 223–44.

Brown, C. and Hegarty, K. (2024) 'Fear and distress: How can we measure the impact of technology-facilitated abuse in relationships?', *Social Sciences*, 13(1): 71–85.

Burrell, S., Ruxton, S., and Westmarland, N. (2019) 'Changing gender norms: Engaging with men and boys', *Government Equalities Office*. Available from: https://www.gov.uk/government/publications/changing-gender-norms-engaging-with-men-and-boys/changing-gender-norms-engaging-with-men-and-boys.

Catton, A.K., Dorahy, M.J., and Yogeeswaran, K. (2023) 'Disclosure of sexual victimization: Effects of invalidation and shame on re-disclosure', *Journal of Interpersonal Violence*, 38(13–14): 8332–56.

Cavanagh, K. and Lewis, R. (1996) 'Interviewing violent men: Challenge or compromise', in K. Cavanagh and V.E. Cree (eds) *Working with Men: Feminism and Social Work*, Routledge, pp 87–112.

Chekola, M. (1994) 'Outing, truth-telling, and the shame of the closet', *Journal of Homosexuality*, 27(3–4): 67–90.

Chen, J.K. (2020) 'Cyber victimisation, social support, and psychological distress among junior high school students in Taiwan and Mainland China', *Asia Pacific Journal of Social Work and Development*, 30(3): 150–63.

Chen, M., Cheung, A.S.Y., and Chan, K.L. (2019) 'Doxing: What adolescents look for and their intentions', *International Journal of Environmental Research and Public Health*, 16(2): 218–32.

Cheung, A. (2021) 'Doxing and the challenge to legal regulation: When personal data become a weapon', in J. Bailey, A. Flynn and N. Henry (eds) *The Emerald International Handbook of Technology-facilitated Violence and Abuse*, Emerald Publishing, pp 577–94.

Christianson, M., Teiler, Å., and Eriksson, C. (2021) '"A woman's honor tumbles down on all of us in the family, but a man's honor is only his": Young women's experiences of patriarchal chastity norms', *International Journal of Qualitative Studies on Health and Well-being*, 16(1): 1–14.

Cinnamon, J. (2017) 'Social injustice in surveillance capitalism', *Surveillance & Society*, 15(5): 609–25.

Clark, H. (2010) '"What is the justice system willing to offer?" Understanding sexual assault victim/survivors' criminal justice needs', *Family Matters*, 85: 28–37.

Clawson, R. (2020) 'Forced marriage of men with learning disabilities: Risks, reasons and responses', in M. Idriss (ed) *Men, Masculinities and Honour-based Abuse*, Routledge, pp 79–94.

Clifford, G., Hitchcock, C., and Dalgleish, T. (2020) 'Compartmentalization of self-representations in female survivors of sexual abuse and assault, with posttraumatic stress disorder (PTSD)', *Psychological Medicine*, 50(6): 956–63.

Coad, D.T. (2017) '"That's my face to the whole field!": Graduate students' professional identity-building through Twitter at a writing studies conference', *Computers and Composition*, 45: 51–66.

Cohen-Almagor, R. and Trottier, D. (2022, March) 'Internet crime enabling: Stalking and cyberstalking', in *Future of Information and Communication Conference*, Springer International Publishing, pp 843–59.

Collins, V.E. and Rothe, D.L. (2023) 'A legacy of harm: Violence against women and the Trump administration', in A. Akande (ed) *US Democracy in Danger: The American Political System Under Assault*, Springer Nature Switzerland, pp 311–30.

Colliver, B. (2021) '"Not the right kind of woman": Transgender women's experiences of transphobic hate crime and trans-misogyny', in I. Zempi and J. Smith (eds) *Misogyny as Hate Crime*, Routledge, pp 213–27.

Cook, E.A. and Walklate, S. (2019) 'Excavating victim stories: Making sense of agency, suffering and redemption', in J. Fleetwood, L. Presser, S. Sandberg, and T. Ugelvik (eds) *The Emerald Handbook of Narrative Criminology*, Emerald Publishing, pp 239–57.

Commission of Inquiry into Queensland Police Service Responses to Domestic and Family Violence. (2022) *A Call for Change Report*, 21 November. Available from: https://www.qpsdfvinquiry.qld.gov.au/about/assets/commission-of-inquiry-dpsdfv-report.pdf.

Connell, R.W. (2005a) *Masculinities*, Polity Press.

Connell, R.W. (2005b) The social organisation of masculinity', in C. McCannn and S.K. Kim (eds) *Feminist Theory Reader: Local and Global Perspectives*, Routledge, pp 103–11.

Connell, R. (2018) 'Conclusion: Reckoning with gender', in J.W. Messerschmidt, P.Y. Martin, M.A. Messner, and R. Connell (eds) *Gender Reckonings: New Social Theory and Research*, New York University Press, pp 331–46.

Conti-Cook, C. (2020) 'Surveilling the digital abortion diary', *University of Baltimore Law Review*, 50: 1–76.

Corbridge, Å. (2018) 'Responding to doxing in Australia: Towards a right to informational self-determination', *University of South Australia Student Law Review*, 3: 1–28.

Criado, N. and Such, J.M. (2015) 'Implicit contextual integrity in online social networks', *Information Sciences*, 325: 48–69.

Criminal Code Act, 1995 (Commonwealth of Australia).

Cross, C., Holt, T., Powell, A., and Wilson, M. (2021) 'Responding to cybercrime: Perceptions and need of Australian police and the general community', *Report to the Criminology Research Advisory Council Grant: CRG 23/16–17*, Australian Institute of Criminology, pp 2021–08.

Crooks, R. (2022) 'Seeking liberation: Surveillance, datafication, and race', *Surveillance & Society*, 20(4): 413–19.

Cuomo, D. and Dolci, N. (2021) 'New tools, old abuse: Technology-enabled coercive control (TECC)', *Geoforum*, 126: 224–32.

Cybercrimes Act No. 19 of 2020 (South Africa).

Daly, K. and Stubbs, J. (2017) 'Feminist engagement with restorative justice', in M. Morash (ed) *Feminist Theories of Crime*, Routledge, pp 503–22.

de Groot, N.F. (2024) 'A contextual integrity approach to genomic information: What bioethics can learn from big data ethics', *Medicine, Health Care and Philosophy*, 27(3): 367–79.

DeKeseredy, W.S. and Schwartz, M.D. (2005) 'Masculinities and interpersonal violence', in M.S. Kimmel, J. Hearn, and R.W. Connell (eds) *Handbook of Studies on Men and Masculinities*, SAGE Publications, pp 353–66.

DeKeseredy, W.S. (2021) 'Image-based sexual abuse: Social and legal implications', *Current Addiction Reports*, 8: 330–5.

di Silvestro, L. and Iurissevich, C. (2024) 'Contemporary pornography: Deepfake as a form of abuse', *European Public & Social Innovation Review*, 9: 1–20.

Didi-Huberman, G. (2009) *Quand les Images Prennent Position. L'oeil de L'histoire, I*, Editions de Minuit.

Donovan, C., Magić, J., and West, S. (2024) 'Family abuse targeting queer family members: An argument to address problems of visibility in local services and civic life', *Journal of Family Violence*, 39(7): 1337–49.

Douglas, D.M. (2016) 'Doxing: A conceptual analysis', *Ethics and Information Technology*, 18(3): 199–210.

Dragiewicz, M., Burgess, J., Matamoros-Fernández, A., Salter, M., Suzor, N.P., Woodlock, D., and Harris, B. (2018) 'Technology facilitated coercive control: Domestic violence and the competing roles of digital media platforms', *Feminist Media Studies*, 18(4): 609–25.

Dunn, J.L. and Powell-Williams, M. (2007) '"Everybody makes choices": Victim advocates and the social construction of battered women's victimization and agency', *Violence Against Women*, 13(10): 977–1001.

Dunn, S. (2021) 'Is it actually violence? Framing technology-facilitated abuse as violence', in J. Bailey, A. Flynn, and N. Henry (eds) *The Emerald International Handbook of Technology-facilitated Violence and Abuse*, Emerald Publishing, pp 25–45.

Dunn, A. (2025) 'The kids aren't alright: Creating greater protections for the children of family vloggers', *Oklahoma Law Review*, 77(2): 527–45.

Dutt, A. (2020) '"Seeing the unseen": Male victims of forced marriages', in M. Idriss (ed) *Men, Masculinities and Honour-based Abuse*, Routledge, pp 24–43.

Eckert, S. and Metzger-Riftkin, J. (2020) 'Doxxing, privacy and gendered harassment. The shock and normalization of veillance cultures', *Medien & Kommunikationswissenschaft*, 68(3): 273–87.

Evteeva, M., Burges, L., and Gelabert, T.S. (2024) 'Internalized misogyny: The patriarchy inside our heads', *Journal of Integrated Social Sciences*, 14(1): 82–108.

Eisenman, T. (2004) *The Accountable Man: Pursuing Integrity Through Trust and Friendship*, InterVarsity Press.

Elias, J. (2024) 'Zemiology in the age of surveillance capitalism: Toward an understanding of the ultra-fast fashion industry', *British Society of Criminology Online Journal*, 22: 83–94.

Elias, J. (2025) 'Towards a digital zemiology', *Justice, Power and Resistance*, 8(1): 40–56.

Ellis, J. (2023) *Representation, Resistance and the Digiqueer: Fighting for Recognition in Technocratic Times*, Policy Press.

Farrell, A. (2021) 'It's just a preference: Indigenous LGBTIQ+ peoples and technologically facilitated violence', in A. Powell, A. Flynn, and L. Sugiura (eds) *The Palgrave Handbook of Gendered Violence and Technology*, Palgrave Macmillan, https://doi.org/10.1007/978-3-030-83734-1_17.

Favarel-Garrigues, G. (2020) 'Digital vigilantism and anti-paedophile activism in Russia: Between civic involvement in law enforcement, moral policing and business venture', *Global Crime*, 21(3–4): 306–26.

Fawcett, J. and Buhle Jr, E.L. (1995) 'Using the internet for data collection: An innovative electronic strategy', *Computers In Nursing*, 13(6): 273–79.

Fawcett, H. and Shrestha, L. (2016) 'Blogging about sexual assault: A thematic analysis', *Journal of Forensic Practice*, 18(1): 39–51.

Feenberg, A. (2002) *Transforming Technology: A Critical Theory Revisited*, Oxford University Press.

Fileborn, B. (2014) 'Online activism and street harassment: Digital justice or shouting into the ether?', *Griffith Journal of Law & Human Dignity*, 2(1): 32–51.

Fileborn, B. (2017) 'Justice 2.0: Street harassment victims' use of social media and online activism as sites of informal justice', *The British Journal of Criminology*, 57(6): 1482–501.

Fileborn, B. and Ball, M. (2023) 'Technology-facilitated domestic violence: Some queer considerations', in B. Harris and D. Woodlock (eds) *Technology and Domestic and Family Violence*, Routledge, pp 133–43.

Finkenauer, C. and Rimé, B. (1998) 'Keeping emotional memories secret: Health and subjective well-being when emotions are not shared', *Journal of Health Psychology*, 3(1): 47–58.

Floridi, L. (2011) 'The informational nature of personal identity', *Minds and Machines*, 21: 549–66.

Floridi, L. (2013) *The Ethics of Information*, Oxford University Press.

Floridi, L. (2017) 'Infraethics: On the conditions of possibility of morality', *Philosophy & Technology*, 30: 391–4.

Floridi, L., Ess, C., Ganascia, J.B., Hildebrandt, M., Laouris, Y., Simon, J., and Thorseth, M. (2015) *The Onlife Manifesto: Being Human in a Hyperconnected Era*, SpringerOpen.

Flynn, A., Powell, A. and Hindes, S. (2023) 'Policing technology-facilitated abuse', *Policing and Society*, 33(5): 575–92.

Flynn, A., Powell, A., Scott, A.J., and Cama, E. (2022) 'Deepfakes and digitally altered imagery abuse: A cross-country exploration of an emerging form of image-based sexual abuse', *The British Journal of Criminology*, 62(6): 1341–58.

Fordyce, R. and Jethani, S. (2021) 'Critical data provenance as a methodology for studying how language conceals data ethics', *Continuum*, 35(5): 775–87.

Foster, S. and Cross, J. (2024) 'Dark doxxing: How dark triad traits impact support for doxxing behaviors', *Personality and Individual Differences*, 217: 1–6.

Fraser, H. and MacDougall, C. (2016) 'Doing narrative feminist research: Intersections and challenges', *Qualitative Social Work*, 16(2): 240–54.

Freeman, G. and Maloney, D. (2021) 'Body, avatar, and me: The presentation and perception of self in social virtual reality', *Proceedings of the ACM on Human-computer Interaction*, 4(CSCW3), pp 1–27.

Frisk, K. (2019) 'What makes a hero? Theorising the social structuring of heroism', *Sociology*, 53(1): 87–103.

Garrick, J. and Buck, M. (2022) 'Doxxing', in M. Buck and J. Garrick (eds) *The Psychosocial Impacts of Whistleblower Retaliation: Shattering Employee Resilience and the Workplace Promise*, Springer International Publishing, pp 153–63.

Gavison, R. (1980) 'Privacy and the limits of law', *The Yale Law Journal*, 89(3): 421–71.

Geary, M. (2021) 'IBAC investigates after police allegedly share photo of former AFL coach Dani Laidley with "vile" comments', *The Guardian*, 10 November. Available from: https://www.abc.net.au/news/2021-11-10/danielle-laidley-ibac-appeal-police-photo-allegation/100610762.

General Data Protection Regulation 2016 (European Union).

Ghumkhor, S. (2019) *The Political Psychology of the Veil: The Impossible Body*, Springer Nature.

Gibson, C., Olszewski, D., Brigham, N.G., Crowder, A., Butler, K.R.B., Traynor, P., Redmiles, E.M., and Kohno, T. (2024) 'Analyzing the AI nudification application ecosystem', *arXiv preprint arXiv:2411.09751*.

Gjika, A. (2023) *When Rape Goes Viral: Youth and Sexual Assault in the Digital Age*, University of California Press.

Goldfeder, S., Kalodner, H., Reisman, D., and Narayanan, A. (2017) 'When the cookie meets the blockchain: Privacy risks of web payments via cryptocurrencies', *Proceedings on Privacy Enhancing Technologies*, 4: 1–19.

Grabosky, P.N. (2001) 'Virtual criminality: Old wine in new bottles?' *Social & Legal Studies*, 10(2): 243–249.

Green, S., Calverley, A., and O'Leary, N. (2021) 'A new approach for researching victims: The "strength-growth-resilience" framework', *The British Journal of Criminology*, 61(3): 852–71.

Grewal, I. (2013) 'Outsourcing patriarchy: Feminist encounters, transnational mediations and the crime of "honour killings"', *International Feminist Journal of Politics*, 15(1): 1–19.

Gross, H. (1979) *A Theory of Criminal Justice*, Oxford University Press.

Hacker, P. (2018) 'Lessons from IP markets for data markets: On moral rights, property rules, and resale royalties', *Intellectual Property Quarterly*, 1: 45–67.

Hage, G. (2009) *Waiting*, Melbourne University Publishing.

Handyside, S. and Ringrose, J. (2017) 'Snapchat memory and youth digital sexual cultures: Mediated temporality, duration and affect', *Journal of Gender Studies*, 26(3): 347–60.

Haraway, D. (1991) *Simians, Cyborgs, and Women: The Reinvention of Nature*, Routledge.

Harder, S.K. and Hasinoff, A.A. (2021) 'Ashamed of shaming? Stories of managing, deflecting, and acknowledging shame after committing image-based sexual abuse', *First Monday*, 21 March, pp 1–15.

Harassment, Harmful Communications and Related Offences Act (HHCRO), 2020 (Republic of Ireland).

Harris, B.A. and Woodlock, D. (2019) 'Digital coercive control: Insights from two landmark domestic violence studies', *The British Journal of Criminology*, 59(3): 530–50.

Harsey, S. and Freyd, J.J. (2020) 'Deny, attack, and reverse victim and offender (DARVO): What is the influence on perceived perpetrator and victim credibility?' *Journal of Aggression, Maltreatment & Trauma*, 29(8): 897–916.

Harmful Digital Communications Act, 2015 (New Zealand).

Hayman, L. (2024) 'What's in a name? Drawing on women's lived experiences to introduce and define cyber-located sexual violence (CLSV)', *Dearcadh: Graduate Journal of Gender, Globalisation and Rights*, 5: 66–84.

Hayman, L.J. (2024) 'Accounting for the harms of image-based sexual abuse in Ireland's Harassment, Harmful Communications and Related Offences Act', *Journal of International Women's Studies*, 26(7): 1–15.

Haywood, C., Johansson, T., Hammarén, N., Herz, M., and Ottemo, A. (2017) *The Conundrum of Masculinity: Hegemony, Homosociality, Homophobia and Heteronormativity*, Routledge.

Hearn, J. (1998) *The Violences of Men: How Men Talk About and How Agencies Respond to Men's Violence to Women*, SAGE Publications.

Heath, M. (2019) 'Espousing patriarchy: Conciliatory masculinity and homosocial femininity in religiously conservative families', *Gender & Society*, 33(6): 888–910.

Heemsbergen, L. (2016) 'From radical transparency to radical disclosure: Reconfiguring (in)voluntary transparency through the management of visibilities', *International Journal of Communication*, 10: 138–51.

Henry, N., Flynn, A., and Powell, A. (2015) 'The promise and paradox of justice: Rape justice beyond the criminal law', in A. Powell, N. Henry and A. Flynn (eds) *Rape Justice: Beyond the Criminal Law*, Palgrave Macmillan UK, pp 1–17.

Henry, N., McGlynn, C., Flynn, A., Johnson, K., Powell, A., and Scott, A.J. (2021) *Image-based Sexual Abuse: A Study on the Causes and Consequences of Non-consensual Nude or Sexual Imagery*, Routledge.

Hoofnagle, C.J., Van Der Sloot, B., and Borgesius, F.Z. (2019) 'The European Union General Data Protection Regulation: What it is and what it means', *Information & Communications Technology Law*, 28(1): 65–98.

Huber, A.R. (2023) 'Image-based sexual abuse: Online communities and the broader misogynistic landscape', *The British Journal of Criminology*, 63(4): 967–83.

Huber, A. and Godfrey, B. (2024) 'Tackling digitally enabled coercive control', *N8 Policing Research Partnership Report*. Available from: https://www.n8prp.org.uk/home/research/small-grants/tackling-digitally-enabled-coercive-control/.

Hussain, N. (1997) 'The narrative appropriation of Saima: Coercion and consent in Muslim Pakistan', in N. Hussain, S. Mumtaz, and R. Saigol (eds) *Engendering the Nation-state*, Simorgh Publications, pp 119–242.

Idriss, M.M. (2020) *Men, Masculinities and Honour-based Abuse*, Routledge.

Idriss, M.M. (2021) 'Abused by the patriarchy: Male victims, masculinity, "honor"-based abuse and forced marriages', *Journal of Interpersonal Violence*, 37(13–14): 11905–32.

Igo, S.E. (2018) *The Known Citizen: A History of Privacy in Modern America*, Harvard University Press.

Information and Electronic Transaction (IN) Law, 2016 (Indonesia).

Iliadis, M., Richards, I., and Wood, M.A. (2020) 'Newsmaking criminology in Australia and New Zealand: Results from a mixed methods study of criminologists' media engagement', *Australian & New Zealand Journal of Criminology*, 53(1): 84–101.

Jane, E.A. (2014) '"Your a ugly, whorish, slut": Understanding e-bile', *Feminist Media Studies*, 14(4): 531–46.

Jane, E.A. (2018) 'Systemic misogyny exposed: Translating rapeglish from the manosphere with a random rape threat generator', *International Journal of Cultural Studies*, 21(6): 661–80.

Jayawickreme, E., Infurna, F.J., Alajak, K., Blackie, L.E., Chopik, W.J., Chung, J.M., and Zonneveld, R. (2021) 'Post-traumatic growth as positive personality change: Challenges, opportunities, and recommendations', *Journal of Personality*, 89(1): 145–65.

Jethani, S. (2015) 'Mediating the body: Technology, politics and epistemologies of self', *Communication, Politics & Culture*, 47(3): 34–43.

Johnson, A. (2005) *The Gender Knot: Unraveling our Patriarchal Legacy* (2nd edn), Temple University Press.

Jørgensen, K.E. and Demant, J. (2021) 'Shame, shaming and economy: A theory of image-based sexual abuse within different online sharing environments', *First Monday*, 26(4–5): 1–15.

Jose, M. (2017) 'International service providers: Mere conduits of data, or gatekeepers to the threshold of cybercrime? What role do ISPs play in ensuring a safe and secure online environment?', *International In-House Counsel Journal*, 10(38): 1–12.

Kadri, T.E. (2020) 'Networks of empathy', *Utah Law Review*, 4: 1075–120.

Käll, J. (2017) 'A posthuman data subject? The right to be forgotten and beyond', *German Law Journal*, 18(5): 1145–62.

Käll, J. (2024) 'More-than-human rights to data', in A. Alvarez-Nakagawa and C. Douzinas (eds) *Non-Human Rights*, Edward Elgar Publishing, pp 189–205.

Kalu, F.A. and Bwalya, J.C. (2017) 'What makes qualitative research good research? An exploratory analysis of critical elements', *International Journal of Social Science Research*, 5(2): 43–56.

Kandiyoti, D. (1988) 'Bargaining with patriarchy', *Gender & Society*, 2(3): 274–90.

Kasra, M. (2017) 'Vigilantism, public shaming, and social media hegemony: The role of digital-networked images in humiliation and sociopolitical control', *The Communication Review*, 20(3): 172–88.

Katz, J. (1988) *Seductions of Crime: Moral and Sensual Attractions in Doing Evil*, Basic Books.

Katz, J. (1995) 'Reconstructing masculinity in the locker room: The Mentors in Violence Prevention Project', *Harvard Educational Review*, 65(2): 163–75.

Katz, S.B. (1989) 'Reviewed work(s): Seductions of crime: Moral and sensual attractions in doing evil', *The Journal of Criminal Law and Criminology*, 80(1): 352–70.

Kay, J.B. and Banet-Weiser, S. (2019) 'Feminist anger and feminist respair', *Feminist Media Studies*, 19(4): 603–09.

Kelly, L. and Westmarland, N. (2016) 'Naming and defining "domestic violence": Lessons from research with violent men', *Feminist Review*, 112(1): 113–27.

Kenney, J.S. and Craig, A. (2012) 'Illegitimate pain: Introducing a concept and a research agenda', in D. Spencer, K. Walby and A. Hunt (eds) *Emotions Matter: A Relational Approach to Emotions*, University of Toronto Press, pp 86–101.

Khan, R. and Lowe, M. (2020) 'Homophobic "honour" abuse experienced by South Asian gay men in England', in M. Idriss (ed) *Men, Masculinities and Honour-based Abuse*, Routledge, pp 95–113.

Khoja-Moolji, S. (2015) 'Becoming an "intimate publics": Exploring the affective intensities of hashtag feminism', *Feminist Media Studies*, 15(2): 347–50.

Kimmel, M. (2013) *Angry White Men: American Masculinity at the End of an Era*, Bold Type Books.

Kitchin, R. and Dodge, M. (2014) *Code/Space: Software and Everyday Life*, MIT Press.

Klucarova, S. and Hasford, J. (2023) 'The oversharenting paradox: When frequent parental sharing negatively affects observers' desire to affiliate with parents', *Current Psychology*, 42(8): 6419–28.

Knowles, C. (2021) 'Beyond adaptive preferences: Rethinking women's complicity in their own subordination', *European Journal of Philosophy*, 30(4): 1317–34.

Kullberg, C. and Skillmark, M. (2017) 'Reluctant help-seekers and agentic victims: Swedish social workers' talk about young men victimised by violence', *Practice*, 29(4): 259–77.

Kumar, A., Hessini, L., and Mitchell, E.M. (2009) 'Conceptualising abortion stigma', *Culture, Health & Sexuality*, 11(6): 625–39.

Kuntsman, A. and Miyake, E. (2022) *Paradoxes of Digital Disengagement: In Search of the Opt-out Button*, University of Westminster Press.

Lamble, S. (2026, in press) 'Carceral diagonalism: The punitive safety politics linking left and right transnational anti-gender mobilisations', *Signs: Journal of Women in Culture and Society*.

Lathan, E. C., Koon-Magnin, S., Selwyn, C.N., Isaak, H. and Langhinrichsen-Rohling, J. (2023) 'Rape myth acceptance and other barriers to formally reporting sexual assault among college students with and without sexual assault histories', *Journal of Interpersonal Violence*, 38(9–10): 6773–97.

Latour, B. (2009) 'Will non-humans be saved? An argument in ecotheology', *Journal of the Royal Anthropological Institute*, 15(3): 459–75.

Lavorgna, A., Ugwudike, P., and Sanchez-Benitez, Y. (2022) 'Harms of digital capital: Social harm analysis of online public resistance and information pollution', *Justice, Power and Resistance*, 5(3): 249–69.

Lee, C. (2022) 'Doxxing as discursive action in a social movement', *Critical Discourse Studies*, 19(3): 326–44.

Legocki, K., Walker, K., and Carriere, T. (2018) 'Doxxing to deter: Citizen activism on social media', *Marketing and Public Policy Conference*. Available from: https://digitalcommons.stmarys-ca.edu/school-economics-business-faculty-works/1274.

Lever, K. and Eckstein, J. (2020) '"I never did those things they said!": Image, coercive control, and intrusion from former partners' technology-mediated abuse', *Iowa Journal of Communication*, 52(1): 49–67.

Lewin, M. (2000) '"I'm not talking to you": Shunning as a form of violence', *Transactional Analysis Journal*, 30(2): 125–31.

Li, Y.T. and Whitworth, K. (2023) 'Data as a weapon: The evolution of Hong Kong protesters' doxing strategies', *Social Science Computer Review*, 41(2): 1650–70.

Lindemann, K. (2010) 'Cleaning up my (father's) mess: Narrative containments of "leaky" masculinities', *Qualitative Inquiry*, 16(1): 29–38.

Lippert, R. (2009) 'Signs of the surveillant assemblage: Privacy regulation, urban CCTV, and governmentality', *Social & Legal Studies*, 18(4): 505–22.

Lomas, T. (2013) 'Critical positive masculinity', *Masculinities & Social Change*, 2(2): 167–93.

Loney-Howes, R. (2018) 'Shifting the rape script: "Coming out" online as a rape victim', *Frontiers: A Journal of Women Studies*, 39(2): 26–57.

Loney-Howes, R. (2020) *Online Anti-rape Activism: Exploring the Politics of the Personal in the Age of Digital Media*, Emerald Publishing.

Lord, K.M. (2006) *The Perils and Promise of Global Transparency: Why the Information Revolution May Not Lead to Security, Democracy, or Peace*, State University of New York Press.

Lorde, A. (1997) [1981] 'The uses of anger', *Women's Studies Quarterly*, 25(1/2): 278–85.

Lupton, D. (2016) 'The diverse domains of quantified selves: Self-tracking modes and dataveillance', *Economy and Society*, 45(1): 101–22.

Lupton, D. (2019) *Data Selves: More-Than-Human Perspectives*, Polity Press.

MacAllister, J.M. (2017) 'The doxing dilemma: Seeking a remedy for the malicious publication of personal information', *Fordham Law Review*, 85(5): 2451–83.

MacKinnon, C.A. (1987) *Feminism Unmodified: Discourses on Life and Law*, Harvard University Press.

Malicious Communications Act, 1998 (United Kingdom).

Marchia, J. and Sommer, J.M. (2017) '(Re)defining heteronormativity', *Sexualities*, 22(3): 267–95.

Massanari, A. (2017) '#Gamergate and the Fappening: How Reddit's algorithm, governance, and culture support toxic technocultures', *New Media & Society*, 19(3): 329–46.

McCuddon, S. (2025) 'Our comment: NPCC guidance advises police to search homes and phones after pregnancy loss', 19 May. Available from: https://www.msichoices.org.uk/news/npcc-advises-police-to-search-homes-and-phones-after-pregnancy-loss/.

McGlynn, C. and Rackley, E. (2017) 'Image-based sexual abuse', *Oxford Journal of Legal Studies*, 37(3): 534–61.

McGlynn, C., Rackley, E., and Houghton, R. (2017) 'Beyond "revenge porn": The continuum of image-based sexual abuse', *Feminist Legal Studies*, 25: 25–46.

McGlynn, C. and Westmarland, N. (2019) 'Kaleidoscopic justice: Sexual violence and victim-survivors' perceptions of justice', *Social & Legal Studies*, 28(2): 179–201.

McGlynn, C., Downes, J., and Westmarland, N. (2017) 'Seeking justice for survivors of sexual violence: Recognition, voice and consequences', in E. Zinsstag and M. Keenan (eds) Restorative Responses to Sexual Violence: Legal, Social and Therapeutic Dimensions. Routledge, pp 179–91.

McIntyre, V. (2016) '"Do(x) you really want to hurt me?": Adapting IIED as a solution to doxing by reshaping intent', *Tulane Journal of Technology & Intellectual Property*, 19: 111–34.

Meyns, S.C. and Dalipi, F. (2022) 'What users tweet on NFTs: Mining Twitter to understand NFT-related concerns using a topic modeling approach', *IEEE Access*, 10: 117658–80.

Migliaccio, T. (2010) 'Men's friendships: Performances of masculinity', *The Journal of Men's Studies*, 17(3): 226–41.

Monreal, T. (2019) '(Re)Learning to teach: Using rasquachismo in the South', *Latino Studies*, 17(4): 118–26.

Mosawi, S.M. (2020) '"You must keep your 'honour': A man without 'honour' is no man at all': Honour, masculinity and violence in Afghanistan', in M. Idriss (ed) *Men, Masculinities and Honour-Based Abuse*, Routledge, pp 128–49.

Nadir, I. and Bakhshi, T. (2018, March) 'Contemporary cybercrime: A taxonomy of ransomware threats & mitigation techniques', in *2018 International Conference on Computing, Mathematics and Engineering Technologies* (iCoMET), IEEE, pp 1–7.

National Police Chiefs' Council (2024) *National Policing Statement 2024 For Violence Against Women and Girls (VAWG)*. https://news.npcc.police.uk/resources/vteb9-ec4cx-7xgru-wufru-5vvo6.

Nelson, S.L. (2024) 'The post-Dobbs reality: Privacy expectations for period-tracking apps in criminal abortion prosecutions', *Pepperdine Law Review*, 51: 783–820.

Nielsen, J. (2003) 'IM, Not IP (Information Pollution) A steady dose of realtime interruptions is toxic to anyone's health', *Queue*, 1(8): 76–5.

Nissenbaum, H. (2004) 'Privacy as contextual integrity', *Washington Law Review*, 79(1): 119–58.

Nissenbaum, H. (2009) *Privacy in Context: Technology, Policy, and the Integrity of Social Life*, Stanford University Press.

Nissenbaum, H. (2018) 'Respecting context to protect privacy: Why meaning matters', *Science and Engineering Ethics*, 24(3): 831–52.

Nissenbaum, H. (2020) 'Protecting privacy in an information age: The problem of privacy in public', in K.W. Miller and M. Taddeo (eds) *The Ethics of Information Technologies*, Routledge, pp 141–78.

Nixon, R. (2011) *Slow Violence and the Environmentalism of the Poor*, Harvard University Press.

OECD. (2024) 'Towards digital safety by design for children', *OECD Digital Economy Papers*, 363, OECD Publishing, https://doi.org/10.1787/c167b650-en.

Ofcom. (2024) *Statement of Strategic Priorities*, 20 November. Available from: https://www.gov.uk/government/publications/draft-statement-of-strategic-priorities-for-online-safety/draft-statement-of-strategic-priorities-for-online-safety.

Öhman, C. (2023) 'How to critique the GDPR: When data protection is turned against the working class', in S. Lindgren (ed) *Handbook of Critical Studies of Artificial Intelligence*, Edward Elgar Publishing, pp 435–44.

O'Neill, T. (2018) '"Today I speak": Exploring how victim-survivors use Reddit', *International Journal for Crime, Justice and Social Democracy*, 7(1): 44–59.

O'Neill, T. (2022) 'Speaking out, "speaking in" and safety work in the digital sphere: Understanding online disclosures in the aftermath of sexual violence', *Journal of Gender-Based Violence*, 6(3): 567–80.

O'Neill, T. (2024) *Disclosing Sexual Violence in a Digital Society: Storytelling, Activism and Justice*, Springer Nature.

O'Neill, T. and Fileborn, B. (2025) ' "These men are fucking disgusting losers": Theorising the affective economies of digital justice and activism', *The British Journal of Criminology*, in press.

Online Safety Act: An Act relating to online safety for Australians, and for other purposes, 2021 (Commonwealth of Australia).

Ortner, S.B. (2022) 'Patriarchy', *Feminist Anthropology*, 3(2): 307–14.

O'Shea, B., Feicht, R., Brown, M., and Numer, M. (2024) 'Rethinking sexual violence labels: Exploring the impact of "victim" and "survivor" discourse', *European Journal of Psychotraumatology*, 15(1): 2296329.

Ozyegin, G. (2018) 'Rethinking patriarchy through unpatriarchal male desires', in J. Messerschmidt, P.Y. Martin, M. Messner, and R. Connell (eds) *Gender Reckonings: New Social Theory and Research*, New York University Press, pp 233–53.

Panfil, V.R. and Miller, J. (2015) 'Feminist and queer perspectives on qualitative methods', in H. Copes and J. Miller (eds) *The Routledge Handbook of Qualitative Criminology*, Routledge, pp 32–48.

Pavan, E. (2017) 'Internet intermediaries and online gender-based violence', in M. Segrave and L. Vitis (eds) *Gender, Technology and Violence*, Routledge, pp 62–78.

Pavan, E. and Lavorgna, A. (2021) 'Promises and pitfalls of legal responses to image-based sexual abuse: Critical insights from the Italian case', in A. Powell, A. Flynn and L. Sugiura (eds) *The Palgrave Handbook of Gendered Violence and Technology*, Palgrave Macmillan, pp 545–64.

Pedersen, W., Bakken, A., Stefansen, K., and von Soest, T. (2023) 'Sexual victimization in the digital age: A population-based study of physical and image-based sexual abuse among adolescents', *Archives of Sexual Behavior*, 52(1): 399–410.

Peters, M.A. (2020) 'Posthumanism, platform ontologies and the "wounds of modern subjectivity"', *Educational Philosophy and Theory*, 52(6): 579–85.

Postma, D. (2016) 'The ethics of becoming in a pedagogy for social justice. A posthumanist perspective', *South African Journal of Higher Education*, 30(3): 310–28.

Powell, A. and Henry, N. (2014) 'Blurred lines? Responding to "sexting" and gender-based violence among young people', *Children Australia*, 39(2): 119–24.

Powell, A. and Henry, N. (2017) 'Sexual violence and harassment in the digital era', in A. Deckert and R. Sarre (eds) *The Palgrave Handbook of Australian and New Zealand Criminology, Crime and Justice*, Palgrave Macmillan, pp 205–20.

Powell, A., Scott, A. J., Flynn, A., and Henry, N. (2020) *Image-based Sexual Abuse: An International Study of Victims and Perpetrators – A Summary Report*, Royal Melbourne Institute of Technology.

Powell, A., Scott, A.J., Flynn, A., and McCook, S. (2022) 'Perpetration of image-based sexual abuse: Extent, nature and correlates in a multi-country sample', *Journal of Interpersonal Violence*, 37(23–24): NP22864–NP22889.

Privacy and Other Legislation Amendment Bill, 2024 (Commonwealth of Australia).

Protection from Harassment Act, 1997 (United Kingdom).

Ralph, B. and Roberts, S. (2020) 'Theories of men and masculinity, and their ability to account for positive change', in R. Magrath, J. Cleland and E. Anderson (eds) *The Palgrave Handbook of Masculinity and Sport*, Palgrave Macmillan, pp 19–38.

Ramsay, P. (2011) 'Preparation offences, security interests, political freedom', in R.A. Duff, L. Farmer, S.E. Marshall, M. Renzo, and V. Tadros (eds) *The Structures of the Criminal Law*, Oxford University Press, pp 203–28.

Ringrose, J., Regehr, K., and Milne, B. (2021) 'Understanding and combatting youth experiences of image-based sexual harassment and abuse', *Department of Education, Practice and Society, UCL Institute of Education.*

Roberts, D.E. (2000) 'The paradox of silence: Some questions about silence as resistance', *University of Michigan Journal of Law Reform*, 33: 343–57.

Rodotà, S. (2009) 'Data protection as a fundamental right', in S. Gutwirth, Y. Poullet, P. De Hert, C. de Terwangne and S. Nouwt (eds) *Reinventing Data Protection?*, Springer Netherlands, pp 77–82.

Romania, V. (2022) 'Shameful traces and image-based sexual abuse: The case of Tiziana Cantone', in F. Comunello, F. Martire and L. Sabetta (eds) *What People Leave Behind: Marks, Traces, Footprints and Their Relevance to Knowledge Society*, SpringerOpen, pp 347–59.

Rosida, I., Ghazali, M.M., Dedi, D., and Salsabila, F.S. (2022) 'The manifestation of internalized sexism in the pick me girl trend on TikTok', *Alphabet: A Biannual Academic Journal on Language, Literary, and Cultural Studies*, 5(1): 8–19.

Røseth, I., Sommerseth, E., Lyberg, A., Sandvik, B.M., and Dahl, B. (2024) 'No one needs to know! Medical abortion: Secrecy, shame, and emotional distancing', *Health Care for Women International*, 45(1): 67–85.

Rouvroy, A. (2013) 'The end(s) of critique: Data behaviourism versus due process', in M. Hildebrandt and K. de Vries (eds) *Privacy, Due Process and the Computational Turn: The Philosophy of Law Meets the Philosophy of Technology*, Taylor & Francis, pp 143–67.

Rowlands, E. (2022) 'Hegemonic masculinity and male powerlessness: A reflection on African men's experiences of intimate partner violence', *South African Review of Sociology*, 52(1): 24–39.

Rutter, N. (2023) 'Rupture, repair, the loss and re-construction of identity: Seeking support in situations of adolescent-to-parent violence and abuse', *Frontiers in Health Services*, 3: 1–7.

Sage, R.A. and Johnson, M.K. (2012) 'Extending and expanding parenthood: Parental support to young adult children', *Sociology Compass*, 6(3): 256–70.

Scharfman, J. (2023) 'Non-fungible token (NFT) fraud', in J. Scharfman (ed) *The Cryptocurrency and Digital Asset Fraud Casebook*, Springer International Publishing, pp 69–80.

Scharp, K.M. and Hall, E.D. (2017) 'Family marginalization, alienation, and estrangement: questioning the nonvoluntary status of family relationships', *Annals of the International Communication Association*, 41(1): 28–45.

Schippers, M. (2007) 'Recovering the feminine other: Masculinity, femininity, and gender hegemony', *Theory and Society*, 36(1): 85–102.

Schneider, E.M. (1991) 'The violence of privacy', *Connecticut Law Review*, 23(1): 973–99.

Schoenebeck, S., Lampe, C., and Triệu, P. (2023) 'Online harassment: Assessing harms and remedies', *Social Media + Society*, 9(1): 1–12.

Schriner, J. and Rood, M.L. (2021) 'The internet never forgets: Image-based sexual abuse and the workplace', in L.R. Salazar (ed) *Handbook of Research on Cyberbullying and Online Harassment in the Workplace*, Investment Company Institute Global Research and Publications, pp 107–28.

Seidler, V.J.J. (1997) *Man Enough: Embodying Masculinities*, SAGE Publications.

Smee, B. (2021) 'Queensland police officer who leaved victim's address resigns', *The Guardian*, 4 October. Available from: https://www.theguardian.com/australia-news/2021/oct/04/queensland-police-officer-neil-punchard-who-leaked-domestic-abuse-victims-address-resigns.

Spencer, D.C. (2015) 'Corporeal realism and victimology', *International Review of Victimology*, 21(1): 31–44.

Spinoza, B. (1992) *Ethics: With the Treatise on the Emendation of the Intellect and Selected Letters*, Hackett Publishing.

Spivak, G. (1994) 'Can the subaltern speak?', in P. Williams and L. Chrisman (eds) *Colonial Discourse in Postcolonial Theory: A Reader*, Columbia University Press, pp 66–111.

Stäcker, D, Franz, A., and Hett, J. (2025) 'Opening Pandora's dox: Investigating dynamics among doxing actors within online environments', *ECIS 2025 Proceedings*, 6: 1–16.

Steinberg, S.B. (2017) 'Sharenting: Children's privacy in the age of social media', *Emory Law Journal*, 66(4): 839–84.

Stephenson, S., Ramjit, L., Ristenpart, T., and Dell, N. (2025, April) 'Digital technologies and human trafficking: Combating coercive control and navigating digital autonomy, in *Proceedings of the 2025 CHI Conference on Human Factors in Computing Systems*, pp 1–21.

Stewart, B. (2016) 'Collapsed publics: Orality, literacy, and vulnerability in academic Twitter', *Journal of Applied Social Theory*, 1(1): 61–86.

Stubbs, J. (2007) 'Beyond apology? Domestic violence and critical questions for restorative justice', *Criminology & Criminal Justice*, 7(2): 169–87.

Sugiura, L. (2021) *The Incel Rebellion: The Rise of the Manosphere and the Virtual War Against Women*, Emerald Publishing.

Summary Offences Act 1966 (Vic) (Commonwealth of Australia).

Suzor, N., Dragiewicz, M., Harris, B., Gillett, R., Burgess, J., and Van Geelen, T. (2019) 'Human rights by design: The responsibilities of social media platforms to address gender-based violence online', *Policy & Internet*, 11(1): 84–103.

Suzor, N., Seignior, B., and Singleton, J. (2017) 'Non-consensual porn and the responsibilities of online intermediaries', *Melbourne University Law Review*, 40(3): 1057–97.

Swartz, T.T., Kim, M., Uno, M., Mortimer, J., and O'Brien, K.B. (2011) 'Safety nets and scaffolds: Parental support in the transition to adulthood', *Journal of Marriage and Family*, 73(2): 414–29.

Sykes, I. (2024) 'From "girlboss" to #stayathomegirlfriend: The romanticisation of domestic labour on TikTok', *European Journal of Cultural Studies*: 13675494241285643.

Szakolczai, J. (2025) 'Onlife criminology: Virtual crimes and real harms', in *Onlife Criminology*, Bristol University Press.

Taylor, J. (2020) *Why Women Are Blamed for Everything: Exposing the Culture of Victim-blaming*, Hachette UK.

The Personal Data (Privacy) Amendment Ordinance, 2021 (Hong Kong).

Tandoc Jr, E.C., Tan Hui Ru, B., Lee Huei, G., Min Qi Charlyn, N., Chua, R.A., and Goh, Z.H. (2022) '#CancelCulture: Examining definitions and motivations', *New Media & Society*, 26(4): 1–19.

Tanzer, D. (1985) 'Real men don't eat strong women: The virgin-madonna-whore complex updated', *The Journal of Psychohistory*, 12(4): 487–95.

Tedeschi, R.G., Calhoun, L.G., and Groleau J.M. (2014) 'Clinical applications of posttraumatic growth', in S. Joseph (ed) *Positive Psychology in Practice: Promoting Human Flourishing in Work, Health, Education and Everyday Life*, Wiley, pp 503–14.

Tegmark, M. (2018) *Life 3.0: Being Human in the Age of Artificial Intelligence*, Vintage Books.

Thompson, C. (2020) 'Skirting around the issue: Misdirection and linguistic avoidance in parliamentary discourses on upskirting', *Violence Against Women*, 26(11): 1403–22.

Thorburn, J., Powell, A., and Chambers, P. (2023) 'A world alone: Masculinities, humiliation and aggrieved entitlement on an incel forum', *The British Journal of Criminology*, 63(1): 238–54.

Tomsen, S. (2017) 'Masculinity and homophobic violence in Australia's recent past', *Sexuality & Culture*, 21: 813–29.

Tomsen, S. and Gadd, D. (2019) 'Beyond honour and achieved hegemony: Violence and the everyday masculinities of young men', *International Journal for Crime, Justice and Social Democracy*, 8(2): 17–30.

Tranchese, A. and Sugiura, L. (2021) '"I don't hate all women, just those stuck-up bitches": How incels and mainstream pornography speak the same extreme language of misogyny', *Violence Against Women*, 27(14): 2709–34.

Trottier, D. (2019) 'Denunciation and doxing: Towards a conceptual model of digital vigilantism' *Global Crime*, 21(3–4): 196–212.

Ugwudike, P., Roth, S., Lavorgna, A., Middleton, S.E., Djohari, N., Tartari, M., and Mandal, A. (2024) 'Sharenting and social media properties: Exploring vicarious data harms and sociotechnical mitigations. Big Data & Society', 11(1): https://doi.org/10.1177/20539517231219243.

Uldbjerg, S. (2020) 'Defying shame: Shame-relations in digital sexual assault', *Mediekultur: Journal of Media and Communication Research*, 36(67): 100–120.

Uldbjerg, S. (2021a) 'Writing victimhood: A methodological manifesto for researching digital sexual assault', *Kvinder, Køn & Forskning*, 1: 27–39.

Uldbjerg, S. (2021b, March 18) 'The rhythms of shame in digital sexual assault: Rhythmic resistance and the repeated assault', *First Monday*: 1–18.

Uldbjerg, S. (2023) 'Digital sexual assault: Understanding the non-consensual sharing of sexual images', in P. Ali and M.M. Rogers (eds) *Gender-based Violence: A Comprehensive Guide*, Springer International Publishing, pp 527–39.

Vangelisti, A. (2006) 'Foreword: Family communication theories: Variations and challenges', in D.O. Braithwaite and L.A. Baxter (eds) *Engaging Theories in Family Communication: Multiple Perspectives*, SAGE Publications, pp ix–xviii.

Vera-Gray, F. (2017) 'Outlook: Girlhood, agency, and embodied space for action', in B. Formal, H. Mulari, and M. Voipio (eds) *Nordic Girlhoods: New Perspectives and Outlooks*, Palgrave Macmillan, pp 127–35.

Verbeek, P.P. (2020) 'Politicizing postphenomenology', in G. Miller and A. Shew (eds) *Reimagining Philosophy and Technology, Reinventing Ihde*, Springer Nature, pp 141–55.

Vogler, S., Kappel, R., and Mumford, E. (2023) 'Experiences of technology-facilitated abuse among sexual and gender minorities', *Journal of Interpersonal Violence*, 38(19–20): 11290–313.

Walklate, S. (2014) 'Sexual violence against women: Still a controversial issue for victimology?'. *International Review of Victimology*, 20(1): 71–84.

Walklate, S., Fitz-Gibbon, K., Reeves, E., Meyer, S., and McGowan, J. (2022) 'In control, out of control or losing control? Making sense of men's reported experiences of coercive control through the lens of hegemonic masculinity', *Journal of Criminology*, 55(4): 451–67.

Walter, M. and Suina, M. (2023) 'Indigenous data, Indigenous methodologies and Indigenous data sovereignty', in S. Swartz, N. Singal, and M. Arnot (eds) *Educational Research Practice in Southern Contexts*, Oxfordshire Routledge, pp 207–20.

Walter, M., Lovett, R., Maher, B., Williamson, B., Prehn, J., Bodkin-Andrews, G. and Lee, V. (2021) 'Indigenous data sovereignty in the era of big data and open data', *Australian Journal of Social Issues*, 56(2): 143–56.

Werbner, P. (2007) 'Veiled interventions in pure space: Honour, shame and embodied struggles among Muslims in Britain and France', *Theory, Culture & Society*, 24(2): 161–86.

Westergren, A. (2016) 'The data liberation movement: Regulation of clinical sharing in the European Union and the United States', *Houston Journal of International Law*, 38(3): 887–912.

Westin, A.F. (1968) Privacy and Freedom, Atheneum Books.

Westmarland, N. and Bows, H. (2018) *Researching Gender, Violence and Abuse: Theory, Methods, Action*, Routledge.

Westmarland, N. and Burrell, S.R. (2023) '"I'm a red-blooded male": Understanding men's experiences of domestic abuse through a feminist lens', *Criminology & Criminal Justice*: 17488958231210985.

Wetboek van Strafrecht, 1997 (Netherlands).

Wheildon, L.J., True, J., Flynn, A. and Wild, A. (2022) 'The Batty effect: Victim-survivors and domestic and family violence policy change', *Violence Against Women*, 28(6–7): 1684–707.

Wijesekera, P., Baokar, A., Hosseini, A., Egelman, S., Wagner, D., and Beznosov, K. (2015) 'Android permissions remystified: A field study on contextual integrity', in *Proceedings of the 24th USENIX Security Symposium*, pp 499–514.

Wilson, M.P. (2020) *The Politics of Privacy Protection: An Analysis of Resistance to Metadata Retention and Encryption Access Laws* [Doctoral dissertation, Queensland University of Technology]. Queensland University of Technology E-Prints, https://eprints.qut.edu.au/199696/.

Wilson, M., Cross, C., Holt, T., and Powell, A. (2022) 'Police preparedness to respond to cybercrime in Australia: An analysis of individual and organizational capabilities', *Journal of Criminology*, 55(4): 468–94.

Williams, R. (1974) *Television, Technology and Cultural Form*, Fontana Press.

Wolf, G. (2016) 'The quantified self: Reverse engineering', in D. Nafus (ed) *Quantified: Biosensing Technologies in Everyday Life*, MIT Press, pp 67–72.

Wong, Y. (2021) 'Victim and the cyber vigilante: An additional perspective on cyber vigilantism', in M. Khader, W.X.T. Chai, and L.S. Neo (eds) *Introduction to Cyber Forensic Psychology: Understanding the Mind of the Cyber Deviant Perpetrators*, World Scientific Publishing, pp 147–64.

Wood, J. (2024) 'Spectre: Covert research in digital far-right "red zones"', in A. Vaughan, J. Braune, M. Tinsley, and A. Mondon (eds) *The Ethics of Researching the Far Right*, Manchester University Press, pp 240–51.

Wood, M.A. (2025) *How Technologies Harm: A Relational Approach*, Bristol University Press.

Wood, M.A., Pervan, F., Lundberg, K., Mitchell, M. and Wood, J. (2025) 'Doing harm with things: Making sense of intention in harmful actions using technology', *Critical Criminology*, https://doi.org/10.1007/s10612-025-09819-2.

Wood, M.A., Mitchell, M., Pervan, F., Anderson, B., O'Neill, T., Wood, J., and Arpke-Wales, W. (2023) 'Inviting, affording and translating harm: Understanding the role of technological mediation in technology-facilitated violence', *The British Journal of Criminology*, 63(6): 1384–404.

Woodlock, D., McKenzie, M., Western, D., and Harris, B. (2020) 'Technology as a weapon in domestic violence: Responding to digital coercive control', *Australian Social Work*, 73(3): 368–80.

Woolley, J., Iliadis, M., and McMahon, M. (2023) 'Digital coercive control: Barriers to victim/survivors' help-seeking and risk management in Victoria', *Journal of Gender-Based Violence*, 7(3): 383–98.

Yar, M. (2005) 'The novelty of 'cybercrime': An assessment in light of routine activity theory', *European Journal of Criminology*, 2(4): 407–27.

Yar, M. and Drew, J. (2019) 'Image-based abuse, non-consensual pornography, revenge porn: A study of criminalization and crime prevention in Australia and England & Wales', *International Journal of Cyber Criminology*, 13(2): 578–94.

Ybarra-Frausto, T. (1989) 'Rasquachismo: A Chicano sensibility', in T. Ybarra-Frausto, L. Eller, and R. Guglielmo (eds) *Chicano Aesthetics*, MARS, Movimiento Artístico del Rio Salado, pp 5–8.

Youngs, G. (2005) 'Ethics of access: Globalization, feminism and information society', *Journal of Global Ethics*, 1(1): 69–84.

Zimmer, M. (2018) 'Addressing conceptual gaps in big data research ethics: An application of contextual integrity', *Social Media + Society*, 4(2): 1–11.

Zuboff, S. (2019) *The Age of Surveillance Capitalism: The Fight For a Human Future at the New Frontier of Power*, Profile Books Ltd.

Zuboff, S. (2022) 'Surveillance capitalism or democracy? The death match of institutional orders and the politics of knowledge in our information civilization', *Organization Theory*, 3(3): 1–79.

Zuboff, S. (2023) 'The age of surveillance capitalism', in W. Longhofer and D. Winchester (eds) *Social Theory Re-Wired*, Routledge, pp 203–13.

Index

A

abortion 37–8, 43–4
abuse allegations 68–9
abuse perpetrators 69–70
accountability 69, 86, 87–8
 see also responsibility
activism 9
affirmative ethics 27–9, 108–9
affirmative model of informational autonomy 111–13
agency *see* resistance to harms
aggrieved entitlement 69
Ahmed, S. 41, 42, 55, 84, 86, 90, 91, 93, 97
AI *see* artificial intelligence
'Aisha' 36, 40, 55–6
Albury, K. 56
alienation from family 53–4
Anderson, B. 5, 21
Anderson, P.S. 86
'Andrew' 68, 73–4
anger 42, 86–7
anonymity 23, 24–5
anxiety 39, 92
apologies 86, 87, 88, 98
appropriateness 25–6
arbitrary data 24–5
artificial intelligence (AI) 11, 33, 34
Australia 5, 57, 58–9, 105–7
autonomy *see* informational autonomy

B

Bailey, J. 7
Banet-Weiser, S. 42
Barnwell, A. 54
Barrett, T. 63
Bennett, J. 21
Bhagawati, M. 90
blackmail 50
blame
 self-blame/guilt 35, 40, 53, 55, 65
 shifting 34, 68, 69, 71
 victim-blaming 34, 69, 71
 see also accountability
Boehnlein, T. 51
Braidotti, R. 27–8, 29, 77, 108
Brison, S.J. 31–2, 85, 95–6
Brown, S. 6
Burges, L. 41
Bwalya, J.C. 107

C

Calhoun, L.G. 110
capitalism 19–20, 23
care 28, 29, 54
 see also support for victim–survivors
'Carol' 50, 104
Catton, A.K. 72
Cavanagh, K. 70
'Chelsea' 90, 94–5, 96, 110–11
Cheung, A. 9
children
 digital rights of 52, 105
 as victims of doxxing 1–2, 52–3, 104
Christianson, M. 56
cisgendered men/women 60–1
Clark, H. 85
coercive control 69, 103
community building 89, 90–2, 95
community participation
 impact of doxxing on 35–7, 40
 sharing experiences 89, 90–2
 withdrawal from 84–5
complaint 89–90, 91, 94
 see also reporting of doxxing
confrontation 86–8, 98
Connell, R.W. 63, 72, 78
conservative/religious values 36–7, 55–6, 73–4
context, link with harm 66
contextual autonomy 112
contextual integrity 25–6, 81
contingent support 54, 73–4
control 69, 103
 loss of 67–8
Corbridge, Å. 5
creativity, following doxing 109–12
Criminal Code Act (1995) 58–9
criminal justice *see* legal responses and remedies

Criminal Law Consolidation Act (1935) 5
criminological framings of doxxing 7–10
cultural values 56

D

DARVO tactics 69, 71
data extraction 26, 28
data leaks
  'Carol' 50, 104
  'Chelsea' 90, 94
data liberation 111
data protection rights 6
data provenance 95
'David' 67–9, 73
DeKeseredy, W.S. 103
Didi-Huberman, G. 109–10
digilantism 9, 71
digital coercive control 69, 103
digital disengagement 83, 94–5, 112
digital evidence 106
digital participation
  as compulsion/necessity 10, 20–1, 83
  disengagement from 83, 94–5, 112
digital spaces 7–8, 19–22, 34–5
  harms of *see* harm(s)
  subjectivity in 18–22
digital subjectivity 18–22
digital technologies
  role in mediating doxxing 104
  see also human–technology relations
digital vigilantism *see* digilantism
Dorahy, M.J. 72
Douglas, D.M. 7
doxxing
  concept 2, 3–4, 101
  creativity and liberation after 109–12
  criminological and sociological framings 7–10
  as form of justice 9
  by police 106–7
  research gaps 12
  research implications 102–8
  responses to *see* family; legal responses and remedies; police; resistance to harms; support for victim–survivors
  types 7, 65
  see also gender differences; harm(s); motivations for doxxing
Dunn, S. 8

E

Eckert, S. 4
Eckstein, J. 103
'Eden' 89, 90–1, 110
'Eila' 35, 89
Elias, J. 40
Ellis, J. 8
emotional responses, working through 95–6
employment *see* work/employment
endurance 85–6
'Ethan' 105
European Union 6
evidence 85, 106
Evteeva, M. 41

F

Facebook 37, 84
faltering autonomy 92
family/ies 49
  alienation from 53–4
  and hierarchy of worth 76
  relational ruptures 37–8, 42, 49, 74
  secrecy in 55, 76
  support from 37, 54–5, 65, 73–5, 77–8
  values and honour of 55–6, 73–4
  see also children
fear 1, 51, 92
feminism 18, 56
Fileborn, B. 41
first-order harms 3, 10, 39
Floridi, L. 11, 22–3, 24–5, 28, 91, 108, 110
Flynn, A. 7
Fordyce, R. 95
forgiveness 86–8, 98
Franz, A. 5
freedom of speech 5
friends, doxxing by 37–8, 44, 48–9, 66, 70–1, 87, 103–4

G

Gadd, D. 67–8
GDPR (General Data Protection Regulation) 6
Gelabert, T.S. 41
gender differences, in doxxing experiences 62, 64, 65, 71–2, 74, 77–8, 90
gender diverse people 8, 43, 107
gender stereotypes/standards 36–7
gendered hierarchies *see* hierarchy of worth
gendered relations
  in digital spaces 7–8
  see also patriarchal context
General Data Protection Regulation (GDPR) 6
Ghumkhor, S. 36
Grabosky, P.N. 11
Groleau, J.M. 110
guilt *see* self-blame/guilt

H

hacked accounts 54–5, 84
hacking (large scale) 90
Hage, G. 85
'Hara' 57, 84–5
harassment 3, 4, 33, 38–9, 51, 52–3
  legal recognition 5, 58–9
  resistance to 83, 84, 86, 92, 97–8

Harassment, Harmful Communications and Related Offences Act (HHCRO) 58
harm(s)
doxxing as harm experience 3–4, 10, 44–5, 104–5
first- and second-order 3–4, 10, 38–9, 77, 80
increased concern about 10–11
see also resistance to harms; self-identity; shame/shaming; social/relational ruptures
Hayman, L.J. 58
Heemsbergen, L. 20
hegemonic femininity/masculinity 49
Henry, N. 7, 11, 48
heterosexual relationships 60–1
Hett, J. 5
HHCRO (Harassment, Harmful Communications and Related Offences Act) 58
hierarchy of worth 32, 34, 36, 49, 63, 67, 76–7, 84
honour 63–4
and family 56, 73–4
humiliation and male 65–7
and respect 68
restoration of 72–6, 77
honour-based violence 63–4
Houghton, R. 56
human–technology relations 10, 20, 21–2, 27, 28
humiliation 64, 65–7, 68, 71

I

identity abuse 39
Idriss, M.M. 72
image-based doxxing 33–4, 36–7, 39–41, 47–9
'Aisha' 36, 40, 55–6
'Andrew' 68, 73–4
'Eila' 35, 89
'Jessie' 54–5, 83, 84, 96–7
'Joe' 66–7, 74–5, 87–8
'Odette' 33–4, 35, 39–40, 51, 58, 85–6
'Sabiyah' 86–7, 88
'Tamara' 38–9, 41–2, 47–9, 59, 83, 84, 93–4
image-based sexual abuse (IBSA) 11, 32–4, 35, 60
incidentalism 70
inforgs 22
informational age 22
informational autonomy 4, 29, 80–1, 92, 102, 104–5, 108–9, 112–13
informational flows 25
informational norms 25–6
informational self 93–5
informational self-determination 91–2, 95–7, 98–9
as liberatory 109–12
infosphere 22, 28
Internet intermediaries 34
interpersonal violence 102, 103–4

J

'Jana' 51, 92
Jane, E.A. 8
'Jessie' 54–5, 83, 84, 96–7
Jethani, S. 20, 95
'Joe' 66–7, 74–5, 87–8
judgement 35, 36, 37, 40–1, 47–51
see also shame/shaming
justice
doxxing as means of 9
social media protections 92
workplace-based 51–2, 92
see also legal responses and remedies; police
justice-seeking
informal strategies 82, 85
see also resistance to harms

K

Käll, J. 27, 106, 108–9
Kalu, F.A. 107
Kandiyoti, D. 76, 77
Katz, J. 65, 66
Kay, J.B. 42
Kelly, L. 70
Khoja-Moolji, S. 21
Kullberg, C. 75
Kumar, A. 37
Kuntsman, A. 83, 94

L

Lamble, S. 8
Lavorgna, A. 9, 28
legal responses and remedies 4–6, 56–9, 82, 85, 105–7
see also police
legislation 4–5, 57, 58–9, 105–6
Lever, K. 103
Lewis, R. 70
LGBTQ+ victim–survivors 8–9
see also gender diverse people; trans people
liberation 109–12
lifespan, privacy across 112
Lindemann, K. 79
Lundberg, K. 21

M

MacAllister, J. 5
MacKinnon, C.A. 18
male violence, and honour 63
manifesto 111–13
Marchia, J. 67
masculine friendships 87–8, 98
masculinity
doxxing as affront to 65–6, 67, 72
and strength 72–3, 75

McGlynn, C. 56
McIntyre, V. 5
medical information 26, 37–8, 43–4
mental health 48, 49
metadata 26, 28
methodological approach 13–14, 107
Metzger-Riftkin, J. 4
misogyny 8, 34
  internalized 41, 44
Miyake, E. 83, 94
'Molly' 1–2, 52–4, 76
motivations for doxxing 9, 50–1
  retribution/payback 36, 40, 44, 48–9, 67–71

N

narrative approach 14
Nissenbaum, H. 24, 25–6, 66
Nixon, R. 32

O

'Odette' 33–4, 35, 39–40, 51, 58, 85–6
O'Donnell, S. 21
O'Neill, T. 41
onlife 11, 22–4
Online Safety Act (2023) 4
online space *see* digital spaces
ontological friction 23, 24
opting in/out 112
  see also digital disengagement
ostracization 53–4
Ozyegin, G. 63, 67, 73, 75

P

partners/ex-partners (doxxing by) 103
  see also 'Aisha'; 'Eila'; 'Sabiyah'
patriarchal context, doxxing in 2, 46, 49, 64, 67, 76–7
patriarchal dividends 63, 72, 75
Pavan, E. 28, 34–5
philosophy of information 22
place 19
  see also digital spaces
police
  misconduct 106–7
  reporting to 35, 53, 57, 58, 93–4
  responses 56–8, 93–4, 106
  skills training 106
political use of digital spaces 20
political use of doxxing 9
post-traumatic growth 110
posthumanism 21, 27–9
'potential onlooker' 48, 50
Powell, A. 11, 48
power 31, 43–4
privacy
  and anonymity 24–5
  contextual integrity of 25–6
  and digital subjectivity 19–21
  gendered use of 18
  in informational age 22–3
  public and 18–19
Privacy Act (1988) 105–6
privacy rights 6, 17–18, 105–6
Priya, K.R. 90
professional reputation 1, 50, 104
publics 18–19

R

Rackley, E. 56
radical disclosure 20
radical healing 90
rape allegations 70–1
relational ruptures *see* social/relational ruptures
religious/conservative values 36–7, 55–6, 73–4
reporting of doxxing 4, 35, 53, 57, 58, 93–4
Republic of Ireland 58
reputation 50, 67–8, 104
resilience 82, 93–5
resistance, doxxing as means of 9
resistance to harms 78–9, 80–2
  community building 89, 90–2, 95
  complaint 89–90, 91, 94
  confrontation and forgiveness 86–8, 98
  creativity and liberation 109–12
  and informational self-determination 95–7, 98–9
  as iterative 92–3
  masculine strength 72–6, 77
  secret-keeping and silence 54–5, 83–4, 86, 97–8
  sharing experiences 89, 90–2, 95
  strength and resilience 93–5
  withdrawal 83–6, 94–5, 97–8
  working through emotional responses 95–6
respect 28, 29
responsibility 34
  see also blame
retribution/payback 36, 40, 44, 48–9, 67–71
rights
  of children 52, 105
  to freedom of speech 5
  to informational autonomy 112–13
  to informational self-determination 95
  to privacy 6, 17–18, 105–6
Romania, V. 36
ruptures
  to self-identity 39–43
  see also social/relational ruptures
Rutter, N. 49

S

'Sabiyah' 86–7, 88
safety/security 1, 38–9, 92
Sanchez-Benitez, Y. 9
Schippers, M. 49

Schneider, E.M. 110, 111
Schwartz, M.D. 103
scrappiness 83, 97
second-order harms 3–4, 10, 38–9, 77, 80
secret-keeping 54–5, 76
Seidler, V.J.J. 73
self-blame/guilt 35, 40, 53, 55, 65
self-determination (informational) 91–2, 95–7, 98–9, 109–12
self-identity 39–43
sense-making 87–8
shame/shaming
  compared with humiliation 65
  in family and community 33–4, 35–7, 49
  gendered nature of 63, 65
  and potential onlooker 48, 50
  and reporting doxxing 35, 58
  stickiness of 41
sharing
  of doxxed information 5, 10, 33, 59, 112
  of victim–survivor experiences 89, 90–2, 95
silence 54–5, 83–4, 86, 97–8
Skillmark, M. 75
slow violence 32, 54, 55
social media
  abuse allegations disclosed on 68–9
  abuse tolerated on 34–5
  account hacking 54–5, 84
  medical information disclosed on 37
  platform liability for sharing 112
  platform protections 92
social networks, doxxing within 65, 68, 71
social/relational ruptures 31, 32–8, 63
  and family responses 37–8, 42, 49, 74
  key impacts of doxxing 76–9
sociological framings of doxxing 7–10
Sommer, J.M. 67
Stäcker, D 5
Steinberg, S.B. 52
Stephenson, S. 105
stickiness 41
stigma 37, 41
strength 72–3, 75, 77, 81–2, 93–5
subjectivity (digital) 18–22
Sugiura, L. 7–8
support for victim–survivors
  as contingent 54, 73–4
  contrasting experiences 53–4, 65
  from families 37, 54–5, 65, 73–5, 77–8
  and hierarchy of worth 76
  male victim-survivors 65, 73–5, 77–8
  see also community building
surveillance 23–4

T

'Tamara' 38–9, 41–2, 47–9, 59, 83, 84, 93–4
Tanzer, D. 76
technology-facilitated violence and abuse (TFVA) 7–9
Tedeschi, R.G. 110
'Tegan' 37–8, 86, 110, 111
telecommunications company data leak 90, 94
timelines of harms 64, 77–8, 84, 85
  see also lifespan; second-order harms
Tomsen, S. 67–8
trans people 8, 43, 107
trauma 31–2
Trottier, D. 9

U

Ugwudike, P. 9
Uldbjerg, S. 48, 49, 50, 51, 94
United Kingdom 4
United States 5

V

Vangelisti, A. 49
Vera-Gray, F. 34, 36
Verbeek, P.P. 20
vibrant materialism 21
victim blaming 34, 69, 71
victim–survivors
  children as 1–2, 52–3, 104
  experiences shared 89, 90–2, 95
  research with 12, 13–14, 70, 107
  see also resistance to harms; support for victim–survivors
violence
  framed as incidental 70
  implications of doxxing research 102, 103–4
  perpetrators 69–70
  shifting blame for 68, 71
viral sharing 5, 10, 33, 59
voice 95, 98
  see also community building; complaint; confrontation; withdrawal

W

'Walter' 70–1
weakness 69
Werbner, P. 37
Westmarland, N. 70
Wilson, M. 57
withdrawal 37, 83–6, 94–5, 97–8, 112
Wood, J. 7

Wood, M.A. 5, 80
work/employment
  loss of job 34, 35, 51
  professional reputation 1, 50, 104
  workplace redress 51–2, 92
worth *see* hierarchy of worth

X

X (social media platform) 92

Y

Yar, M. 11
Ybarra-Frausto, T. 83
Yogeeswaran, K. 72
Youngs, G. 20

Z

Zuboff, S. 10, 106

Printed and bound by CPI Group (UK) Ltd, Croydon, CR0 4YY

15/09/2026

14962386-0008